T0075240

AI in and for Africa

AI in and for Africa: A Humanistic Perspective explores the convoluted intersection of artificial intelligence (AI) with Africa's unique socio-economic realities. This book is the first of its kind to provide a comprehensive overview of how AI is currently being deployed on the African continent.

Given the existence of significant disparities in Africa related to gender, race, labour, and power, the book argues that the continent requires different AI solutions to its problems, ones that are not founded on technological determinism or exclusively on the adoption of Eurocentric or Western-centric worldviews. It embraces a decolonial approach to exploring and addressing issues such as AI's diversity crisis, the absence of ethical policies around AI that are tailor-made for Africa, the ever-widening digital divide, and the ongoing practice of dismissing African knowledge systems in the contexts of AI research and education. Although the book suggests a number of humanistic strategies with the goal of ensuring that Africa does not appropriate AI in a manner that is skewed in favour of a privileged few, it does not support the notion that the continent should simply opt for a "one-size-fits-all" solution either. Rather, in light of Africa's rich diversity, the book embraces the need for plurality within different regions' AI ecosystems. The book advocates that Africa-inclusive AI policies incorporate a relational ethics of care which explicitly addresses how Africa's unique landscape is entwined in an AI ecosystem. The book also works to provide actionable AI tenets that can be incorporated into policy documents that suit Africa's needs.

This book will be of great interest to researchers, students, and readers who wish to critically appraise the different facets of AI in the context of Africa, across many areas that run the gamut from education, gender studies, and linguistics to agriculture, data science, and economics. This book is of special appeal to scholars in disciplines including anthropology, computer science, philosophy, and sociology, to name a few.

Chapman & Hall/CRC
Artificial Intelligence and Robotics Series
Series Editor: Roman Yampolskiy

Autonomous Driving and Advanced Driver-Assistance Systems
Applications, Development, Legal Issues, and Testing
Edited by Lentin Joseph, Amit Kumar Mondal

Digital Afterlife and the Spiritual Realm
Maggi Savin-Baden

A First Course in Aerial Robots and Drones
Yasmina Bestaoui Sebbane

AI by Design
A Plan for Living with Artificial Intelligence
Catriona Campbell

The Global Politics of Artificial Intelligence
Edited by Maurizio Tinnirello

Unity in Embedded System Design and Robotics
A Step-by-Step Guide
Ata Jahangir Moshayedi, Amin Kolahdooz, Liao Liefa

Meaningful Futures with Robots
Designing a New Coexistence
Edited by Judith Dörrenbächer, Marc Hassenzahl, Robin Neuhaus, Ronda Ringfort-Felner

Topological Dynamics in Metamodel Discovery with Artificial Intelligence
From Biomedical to Cosmological Technologies
Ariel Fernández

A Robotic Framework for the Mobile Manipulator
Theory and Application
Nguyen Van Toan and Phan Bui Khoi

AI in and for Africa
A Humanist Perspective
Susan Brokensha, Eduan Kotzé, Burgert A. Senekal

For more information about this series please visit: https://www.routledge.com/Chapman--HallCRC-Artificial-Intelligence-and-Robotics-Series/book-series/ARTILRO

AI in and for Africa
A Humanistic Perspective

Susan Brokensha
Eduan Kotzé
Burgert A. Senekal

CRC Press is an imprint of the
Taylor & Francis Group, an **informa** business
A CHAPMAN & HALL BOOK

Cover image: Shutterstock

First edition published 2023
by CRC Press
6000 Broken Sound Parkway NW, Suite 300, Boca Raton, FL 33487-2742

and by CRC Press
4 Park Square, Milton Park, Abingdon, Oxon, OX14 4RN

CRC Press is an imprint of Taylor & Francis Group, LLC

ISBN: 978-1-032-23176-1 (hbk)
ISBN: 978-1-032-21575-4 (pbk)
ISBN: 978-1-003-27613-5 (ebk)

DOI: 10.1201/9781003276135

Typeset in Sabon
by SPi Technologies India Pvt Ltd (Straive)

Contents

Preface ix
Foreword xi
About the authors xiii

1 Perceiving AI through a humanistic lens **1**

Introduction 1
Humanism and AI 2
Ubuntu *and AI 4*
Decolonial and post-colonial lenses 6
Notes 9

2 Intricate intersections: AI's diversity and governance crises **11**

Introduction 11
The portrayal of AI as white 11
Women in (emerging) technology 14
Algorithmic exploitation 17
Ecology in two senses of the word 20
Ethical AI and policymaking 21
A humanistic view of ethical AI governance 25

3 AI ethics in and for Africa: Some preliminary thoughts **27**

Introduction 27
Disconnects 28
Ethical debates around AI 29
A relational ethics of care 33
Feminist and engaged commitments to AI 42
Notes 43

4 (Mis)perceptions of AI in Africa: Metaphors, myths, and realities 45

Introduction 45
African perceptions of AI 46
Techno-utopia versus techno-dystopia 49
Making (non)sense of AI as a technical artefact 52
Note 56

5 Digital citizenship in Africa: Contestations and innovations in the context of AI 57

Introduction 57
Responses to the curtailment of digital citizen engagement 58
Responses to unequal digital access and (responsible) digital literacy 60
Responses to digital citizenship through digital education and research 62
Promoting digital citizenship in Africa in the context of AI: Moving beyond a symbolic term 65
Notes 69

6 AI and the labour sector in Africa: Disruptive or transformative? 71

Introduction 71
Artificial intelligence and the disruption of the labour market 72
Semi-skilled labour 72
Unskilled labour 73
Skilled labour 73
A pessimistic view of the impact of automation on the labour market 75
An optimistic view of the impact of automation on the labour market 76
AI ethics in the workplace 77
The impact of automation on education and training 78

7 **Machine learning, deep learning, text mining, and computer vision in Africa: Deployments and challenges** 83

Introduction 83
Machine learning 85
Deep learning 86
Text mining 90
Computer vision 92
Note 93

Postface 95

References 97
Index 133

Preface

Resisting exaggerated expectations about the capabilities of artificial intelligence (AI) and acknowledging that this transformative technology may be on the verge of a new winter – or at least an autumnal phase – this book interrogates AI's role in Africa from a humanistic (socio-technical) perspective. It specifically aims to explore this technology's complicated convergences with Africa's socio-economic realities, recognising that how the fourth industrial revolution is evolving on the continent simply does not mirror how it is unfolding in developed regions of the world. Considering the continent's complex gender, race, labour, and power dynamics, among others, the stance adopted is that what is needed are African solutions to such dynamic forces that are not simply Eurocentric or technologically deterministic. Without entirely dismissing Western knowledge systems, a call is made for a decolonial approach to the design and deployment of AI. This does not mean advocating generic AI-driven solutions to African problems: recommending an all-encompassing prescription to remedy such a diverse continent's ills would be brazen indeed and ignores plurality and pluriversality, while also disregarding the continent's colonial era legacies.

From the outset, it is important to note that the book frames AI technologies as narrow or weak – as systems that have been designed to carry out tasks that are singular or limited. Since artificial general intelligence has not been achieved, it would not at this point make sense to portray it as anything but narrow. Although AI's deployment on the continent is not yet ubiquitous, the book stresses that narrow AI is not without its risks and that it is humans who develop and implement it in ways that are not necessarily beneficial to all.

Foreword

The development of artificial intelligence (AI) is inextricably bound up in the relations of power which structure global society. Whereas overenthusiastic tech writing often presents AI systems as developing in a social vacuum, a more careful analysis quickly reveals how hegemonic actors are shaping the way technology is being developed to their own advantage. The gains of AI systems are being distributed much like every other resource: that is to say, along the lines of private property and concentrated ownership. The increasing concentration of the ability to develop AI only threatens to accelerate this trend (Srnicek, 2022). This inequality does not annul the possible benefits attached to AI development – but it should make us think twice about uncritical narratives which present AI technology as a panacea that has the potential to cure all social ills. As Kranzberg (1986) expresses it in his first rule: "technology is neither good nor bad; nor is it neutral" (p. 547).

The asymmetry produced by power does not just play out in the concrete development of AI systems but also in the discourse surrounding that development. The AI ethics debate, for instance, has systematically underplayed the significance of the application of AI to the labour process. Work is a high-risk environment, and it is where many people will encounter AI at close hand for the first time (Lee et al., 2015) – and yet many ethical debates have skirted the issue entirely, and those that have engaged with it have often ignored workers and unions (Cole et al., 2022). Class, as ever, turns out to be one of the fundamental determining relationships that shapes the particular form taken by a social phenomenon.

But the relation between classes does not only internally structure national economies. The contemporary economic system is also, always, a world-system (Arrighi, 2010). Externally, the relations between different national economies are also shaped by class and the structures of power that come alongside it. In *AI in and for Africa*, Brokensha, Kotzé, and Senekal take a decolonial, humanistic perspective on precisely these questions of how AI development is being shaped by social power from the diverse standpoints of Africa's 54 countries and 1.4 billion people.

What they find should give AI's many uncritical proponents pause for thought. Far from being circumnavigated by these new developments, the historical relations of unequal exchange and exploitation between the Global North and Global South are being reanimated by them. The list of issues catalogued by Brokensha et al. is extensive: from exclusion and prejudice along the lines of race and gender to less beneficial or even actively harmful AI system outcomes. They paint a picture of a process of development which is neither by nor for the vast majority of Africa's population.

But at the same time that African actors are excluded from the upper tiers of AI system development and from the populations for which these systems are being developed, they are integrated at the lowest. Enlisted as cloudworkers *en masse* to complete tasks such as labelling the training data that is so vital for machine learning, African workers are asked to act as an exploitable low-wage population (Fairwork, 2021; Posada, 2022). Their (micro)job is to enable a process of development that excludes them at every other level. Rather than reaping the rewards of AI systems in terms of productivity and efficiency, the continent is at risk of bearing the costs: be they those associated with the informal digital labour of cloudworkers or the mining of raw materials like lithium, which are foundational to AI system production networks. Chapter by chapter, Brokensha et al. marshal a huge range of literature to make the risks of this path extremely clear.

In this sense, the accumulation of resources enabled by the colossal violence of imperialist accumulation by dispossession "weighs like a nightmare on the brains of the living" (Marx, 1852, p. 103). The solutions proposed below are broad and insightful: supporting an epistemic reframing of the risks and benefits of AI, advancing a relational ethics of care that addresses Africa's specific contexts, preparing for the transformation of the labour market through regulatory and training changes that support fairness, and more.

AI in and for Africa makes obvious the moral and practical need for a movement to reverse the systematic underdevelopment of the Global South. The kind of social transformation required to achieve substantial change in the distribution of both literal and epistemic resources should not be underestimated. Achieving widespread political support for such change has so far eluded anti-imperialist political actors in the Global North. But this state of failure is not guaranteed to last forever, and it does not lessen that underlying need. In fact, given the existential risk posed to billions of people, primarily located in the Global South, by "capitalogenic" climate change (Moore, 2017), this movement's emergence seems more urgent now than ever before. Scholars of AI, wherever they are located, have a duty to look this crisis in the face and advocate for more than what seems "possible" within the coordinates of our current economic system.

Callum Cant
Oxford Internet Institute, University of Oxford

About the authors

Susan Brokensha is an Applied Linguist at the University of the Free State, South Africa, and co-convenor of the ethics and governance group located in the Interdisciplinary Centre for Digital Futures (ICDF) at the university.

Eduan Kotzé is an Associate Professor and Head of the Department of Computer Science and Informatics at the University of the Free State.

Burgert A. Senekal is a Research Fellow in the Department of Computer Science and Informatics at the University of the Free State.

Chapter 1

Perceiving AI through a humanistic lens

INTRODUCTION

Although a humanistic perspective of AI may appear to inconvenience technology companies (Gill, 2017, p. 475), with some regarding humanism as a catchphrase devoid of meaning (Ostherr, 2018), such a perspective takes into account that social, cultural, ethical, and ecological dimensions are central to the development and application of AI (Ekanem, 2013, p. 81). Indeed, a number of scholars have in recent years either implicitly or explicitly called for a humanistic perspective on novel technologies in Africa and represent a variety of disciplines as diverse as law (Mahomed, 2018), information and communications technology (Nayebare, 2019), medicine (Ewuoso and Fayemi, 2021), and philosophy (Ihejirika, 2015; Lamola, 2021). From the outset, and to avoid unnecessary confusion which may arise when the terms "humanism", "humanist", and "humanistic" are used interchangeably, we partially define a humanistic approach to AI as one that not only interrogates the values and ethical tenets that should underlie AI's design and implementation but also ensures the dignity and well-being of humans and of humanity (cf. Zhou et al., 2020, p. 3011). Additionally, we posit that such a paradigm should continuously evaluate AI's impact on the natural environment, which includes animals. With respect to the latter, environmental and space ethicist Andrea Owe and executive director of the Global Catastrophic Risk Institute Seth Baum note that the vast majority of (draft) documents or guidelines on the ethics of AI currently in circulation rarely consider the intrinsic value of non-human animals, focusing instead on the moral status of AI itself (Owe and Baum, 2021). Our approach thus abjures both AI-centrism and anthropocentrism, calling for a socio-technical view that also takes into account how AI mediates human-ecological relationships (cf. Ahlborg et al., 2019).

Since all facets of AI addressed in this book are examined through a humanistic lens, it goes without saying that the chapter begins with a discussion of humanism from which the adjective is derived. Attention is paid to deconstructing the term in the context of the African philosophy of *ubuntu*, which has its roots in African humanism and whose humanistic tenets

DOI: 10.1201/9781003276135-1

appear to address Africa's concerns about AI's societal, political, economic, and ethical impacts – at least to some extent. What we mean by this is that the choice to explore the design and deployment of AI in and for Africa through the sole lens of *ubuntu* would be a fraught one: not only is it a near-impossible feat to capture exactly what *ubuntu* encompasses, but it cannot be taken for granted that it is in widespread use on the continent (Ewuoso and Hall, 2019, p. 93). What is more, and as we shall see, several significant criticisms have been levelled against it (cf. West, 2014). Although we consider aspects of Western humanism that overlap with those of African humanism (Edeh, 2015), neither worldview offers unassailable solutions to AI's challenges (cf. Pietersen, 2005). To address the deficiencies reflected in the two approaches, we advance a humanistic view of AI in Africa that draws on post-colonial and decolonial theories, in particular Mohamed, Png and Isaac's (2020) socio-technical framework, which to some extent encompasses Jansen's (2017, pp. 156–163) conceptions of decolonisation that include, among others, a focus on decentring Eurocentric knowledge, the addition of African knowledge systems to existing canons[1], and the implementation of an "engagement view" of science (Mohamed et al., 2020, p. 664). As will become clear, a decolonial framework such as this one is useful, as it speaks directly to the benefits and harms of AI in developing countries. It also offers more nuanced perspectives as it avoids both unhelpful binaries "of West and the rest, of scientists and humanist [s]" (Mohamed et al., 2020, p. 676) and *ubuntu's* somewhat vague and varied principles that have, in West's (2014, p. 49) view, culminated in this philosophy being simplistically conceptualised as one that labels values as either "good" or "bad".

Given that humanism is an ambiguous and contested term, what it entails in relation to AI is clarified before we consider Mohamed et al.'s (2020) framework.

HUMANISM AND AI

As Igwe (2002, p. 190) avers, it is challenging to attach a single definition to humanism since it accommodates a host of ideologies and has been variously described as "a 'worldview', an 'approach to life', a 'life stance', an 'attitude', a 'way of life', and a 'meaning frame'" (Copson, 2015, p. 5). In attempting to elucidate this contested term (Lee, 2018, p. xiii), Copson (2015, p. 6) quotes the International Humanist and Ethical Union (IHEU) as defining it as "a democratic and ethical life stance, which affirms that human beings have the right and responsibility to give meaning and shape to their own lives" (Bylaw 5.1 of the IHEU). Although scholars refer to a type of universal humanism that recognises different races, cultures, and traditions (cf. Copson, 2015, p. 2; Svetelj, 2014, p. 23), it remains an approach that is slanted in favour of Eurocentric or Western-centric discourses (cf. Pinn, 2021, p. xviii). This skewedness is particularly evident in the field of AI as it pertains to how knowledge is perceived: African knowledge systems are regularly sidelined by Western systems (cf. Hlalele, 2019, p. 89), a tendency that is apparent in the integration of AI-enabled technology into African healthcare systems: it has, for instance, been noted that AI diagnostic tools for breast cancer developed in high-income countries – and thus trained on mammograms from these countries – are inappropriate for use in Sub-Saharan Africa, where women develop breast cancer at an earlier age and have poorer prognoses (Black and Richmond, 2019; Canales, Lee and Cannesson, 2020). Coupled to this problem, and of increasing concern, is that medical AI has revealed algorithmic bias which may be racial in nature (Obermeyer et al., 2019).

We are not suggesting that Western humanism be entirely negated in discussions of how AI could or should be operationalised on the African continent. However, while this worldview stresses human potential and therefore self-actualisation (as many types of humanism do), its emphasis on a core humanity makes it "culturally specific" (Gaylard, 2004, p. 266), while notions of diversity and heterogeneity are de-emphasised. Further, its Eurocentric stance attempts to provide justifications for the colonial project on the continent (cf. Gaylard, 2004, p. 266). One need look no further than studies on AI-driven algorithms to establish the truth of the latter concern: as Birhane (2020a) argues, many of the algorithmic models currently in use essentially perpetuate colonialism, given that they are not tailor-made for Africa's needs and obstruct opportunities to design and deploy local AI technologies. In Zondi's (2021, p. 229) view, the hegemony of Western technologies is such that we are inclined to overlook the fact that technologies have been generated over centuries by myriad civilisations across the globe. With regard to Africa's contribution to technological development, African legal technology scholar Nathan-Ross Adams attributes AI's sketchy history on the continent to colonisers' attempts to "violently [erase] our ancestors' dreams and thoughts on AI from history" (Adams, 2019, para. 5). Although mainstream accounts of technology tend to dismiss Africa's connections to AI, these connections certainly exist. Siyonbola (2021), for example, offers some intriguing insights into Africa's AI origins, describing divination systems such as the Ifa system of the Yoruba people (Alamu, Aworinde and Isharufe, 2013) as being linked with computer science and which too have a foundation in advanced mathematics. As she points out, Africa's "divination systems have always been based on data and binary systems, which is exactly what Google is: an amalgamation of an insane amount of data being queried by different algorithmic combinations" (Siyonbola, 2021, para. 4).

In attempting to counter significant concerns about AI being Western-centric in its design and implementation, we look to some principles of *ubuntu* that – at least in part – lend themselves well to addressing some of these concerns (Mhlambi, 2020). As indicated earlier, we do not view AI solely from the perspective of *ubuntu*. After all, as Biney (2014, p. 29) reminds us, this philosophy has its origins in southern Africa and bears some similarities to African humanism that emerged in countries across the continent (Biney, 2014, p. 29). Guided by Biney (2014, p. 30), we avoid an essentialist approach, positing that through their lived experiences, communities define *ubuntu* in ways that are uniquely relevant to them. Since *ubuntu* acknowledges that human diversity as it relates to values, customs, and language, for instance, needs to be respected, it is a philosophy that also recognises plurality, the importance of which cannot be overestimated within AI ecosystems: Rudschies, Schneider and Simon (2020, p. 2) declare that while attempts to achieve mutual understandings of AI's ethical principles are all very well, divergent voices are of significant value, exposing conflicting debates around what this technologies' values and principles should entail. Divergence essentially uncovers underlying structures and dynamics of AI's myriad facets that could potentially benefit or harm societies and communities. What is called for then are voices and analyses of difference (Zondi, 2021, p. 240) around AI as espoused by a number of scholars working in the field (Adams, 2021; Mohamed et al., 2020; Rudschies et al., 2020). This plurality, as will become clear, is well-aligned with a humanistic perspective of AI. Before reviewing this perspective, we examine what *ubuntu* is and how this philosophy relates to AI. At the same time, we acknowledge some of the criticisms levelled against it to demonstrate why we have not employed it as the main lens through which to view AI in Africa.

Ubuntu and AI

Whether reference is made to Steve Biko's Black Consciousness, to the philosophy of Ujamaa as proposed by Julius Nyer, to Humanism as championed by Kenneth Kaunda, or to Kwame Nkrumah's Consciencism, to name a few, it would be shortsighted to conclude that the main difference between these forms of African humanism and Western humanism is simply that the former emphasises communalism, while the latter calls for individualism (Gaylard, 2004, p. 279; Naudé, 2019, p. 226). Applied ethicist Motsamai Molefe (2021) argues that while social relationships are important in African ethics, their role has nevertheless been inflated by scholars. We cannot after all dismiss the kind of self-interest that, as opposed to selfish self-interest, Pellegrino (1989, p. 57) rightly claims "pertains to the duties we owe to ourselves – duties which guard health, life, some measure of material well-being, the good of families, friends, etc". African humanism acknowledges the individual but is averse to an individualism that is radical in nature (Obioha and Okaneme, 2017, p. 44), viewing an individual's identity and morality as being intertwined with that of other individuals in society (Winks, 2011, p. 456).[2] In southern Africa, African humanism finds its expression in *ubuntu*, which in a literal sense means "humaneness" and may be defined as "a multidimensional concept representing a core value of African ontologies" (Sambala, Cooper and Manderson, 2020, p. 50). As a concept, it is found in most African countries (cf. Mabvurira, 2020), albeit that its meaning changes according to language (cf. Van Breda, 2019, p. 439). In the Democratic Republic of the Congo, for example, the equivalent of *ubuntu* is *bomoto*, which translates as "humanity", while in Botswana, the equivalent *botho* is a concept that refers to personhood reflecting the belief that an individual "is regarded as a person through other people" (Eleojo, 2014, p. 305).

That there has in recent years been a proliferation of studies calling for *ubuntu* tenets to be incorporated into technology (e.g., Kapuire et al., 2015; Kemeh, 2018; Mhlambi, 2020; Nayebare, 2019) is not unexpected, as it is a philosophy that advocates respect for human dignity in relationships mediated through technology (cf. Gabriel, 2020, p. 424). Mindful that *ubuntu's* universality is questionable (cf. West, 2014, p. 58), we base our humanistic stance towards human–AI interaction on the assertion made by Metz (2007) that "An action is right just insofar as it produces harmony and reduces discord"; an act is wrong to the extent that it fails to develop community" (Metz, 2007, p. 334). We thus partially interrogate AI through the *ubuntu* principles of solidarity, equality, equity, and community as proposed by Mhlambi (2020), who discusses them in the context of what he describes as a framework for the governance of AI that recognises respect for both ethics and human rights.

Although emerging technologies such as AI have engendered myriad possibilities for social good, they also threaten to destabilise social bonds. In Mhlambi's (2020) ethical and human rights framework, it is essential that AI ecosystems preserve solidarity through social cohesion to prevent and challenge human rights abuses, as well as to affect social change. In this respect, Mohamed et al. (2020, p. 676) observe that forms of solidarity geared towards the decolonisation of power in the field of AI include Black in AI, Data for Black Lives (Goyanes, 2018), and the Deep Learning Indaba (Gershgorn, 2019). The fact that solidarity or social cohesion is an important tenet of *ubuntu* implies that much value is also attached to consensus-making within communities, and so it is unsurprising that scholars have pointed out that this contradicts the argument that *ubuntu* respects plurality and individuality. Yet, in addressing the question as to whether or not *ubuntu* has the capacity to accommodate these two notions, Louw (2010, p. 7) replies with a definitive "yes", contending that *ubuntu's* "principle of agreeing to disagree [...] can handle plurality insofar as it inspires us to expose ourselves to others, to encounter the differences of their humanness so as to inform and enrich our own". This is an acknowledgement not only of personal identity, which is perceived as being embedded in reciprocal interconnection, but also of "otherness", which is critical both for embracing plural worldviews referred to earlier and for resisting the kind of *ubuntu* that may repress dialogue (Louw, 2010, p. 4).

Adherence to the second and third tenets, namely equity and equality, is critical if power asymmetries in both the design and implementation of AI in Africa are to be adequately addressed. AI models trained on non-representative or biased datasets, for example, simply exacerbate inequalities on the social and economic dimensions as do the actions of technology companies and unscrupulous governments, who may leverage this technology to breach people's privacy and violate their human rights (Gwagwa et al., 2020) (see Chapter 3). With respect to gender equality, we acknowledge the concern voiced by several scholars that while *ubuntu* was initially conceptualised as "gender blind" (Du Plessis, 2019, p. 44), it does, like other African philosophies, enforce patriarchy (Du Plessis, 2019, p. 44, cf. Manyonganise, 2015; Oyowe and Yurkivska, 2014; Waghid and Smeyers, 2012) first established through coloniality. In Chapters 2 and 3, we briefly consider *ubuntu* feminism as a possible way to address gender disparities in the context of AI.

In explicating the principle of community, Mhlambi (2020) quite rightly points out that it is critical to both recognise and respect vulnerable populations who may be excluded from participating in the design and implementation of AI-enabled technologies. Not including marginalised or vulnerable communities from AI ecosystems is a dismissal, among other things, of recognition of different ethical systems around AI and of respect for the incorporation of indigenous knowledge systems into AI models in a variety of sectors.

Mhlambi's (2020) adherence to these principles is premised on the argument that *ubuntu* emphasises the relational dimension of personhood, and so it resonates with many researchers in Africa, who argue that Western theories pay scant attention to this aspect and accentuate personhood's individualist aspect instead (cf. Wareham, 2020, p. 132). In Naudé's (2019) view, this distinction is rather crude, however, in light of the fact that "In no society, neither Western nor African, can an individual create him- or herself *ex nihilo* or outside of social relations [...]" (Naudé, 2019, p. 226). What is more, Naudé (2019, p. 226) quotes anthropologists John Comaroff and Jean Comaroff (2002, p. 78) as observing that "Nowhere in Africa were ideas of individuality ever absent", which is why we do not accept the notion of relational personhood as being unique to *ubuntu*.

Considering the problems inherent in some of the empirical claims made about *ubuntu*, this philosophy is deployed as an underlying trope throughout this book, while facets of AI in Africa are explored within Mohamed et al.'s (2020) more nuanced framework of structural decolonisation illuminated in the next section.

DECOLONIAL AND POST-COLONIAL LENSES

A discussion of AI on the African continent cannot be divorced from a consideration of how power imbalances and inequities related to this technology are entrenched in colonial exploitation. Whether one is referring to the colonisation of countries that exhibited a centralised state (such as Burundi or Rwanda), to those settled by whites (such as Kenya and Namibia), or to those "which did not experience significant white settlement and where there was either no significant pre-colonial state formation (like Somalia or South Sudan) or where there was a mixture of centralised and uncentralised societies" (Heldring and Robinson, 2013, para. 8), colonisation devasted Africa's autonomy, depleting its natural resources and disarticulating the social, cultural, political, and intellectual lives of its people. As will be discussed in greater detail, colonisers of Africa from the seventeenth century onwards partially justified the colonial project by appeals to the racial and intellectual superiority of white men (Cave and Dihal, 2020, p. 2), harnessing pseudo-sciences such as eugenics and racial science to legitimise these constructions (Benyera, 2021, p. 28). Significantly, the white man's supposed intellectual advantage over "inferior" groups of people extended to technological superiority too (Cave and Dihal, 2020, p. 2), a superiority which, as we will elucidate, is now perpetuated in the era of AI too, which is why, like Mohamed et al. (2020), we argue that discussions of AI in Africa cannot be removed from an acknowledgement of post-colonial abuses and inequalities. Fuelled by the emergence of the fourth industrial revolution, Africa has now entered what Benyera (2021) conceives of as "the second phase of the colonisation of [the continent]" (p. 32).

To interrogate how AI's colonial past has bearing on its current design and deployment in Africa as well as to imagine its future, we are guided by aspects of Mohamed et al.'s (2020) socio-technical approach, an approach that, adapted to the AI landscape, may be defined as one "that calls attention to how values, institutional practices, and inequalities are embedded in AI's code, design, and use" (Joyce et al., 2021, p. 1). Encompassing a broad critical science approach, Mohamed et al.'s (2020) socio-technical lens, as indicated earlier, views structural decolonisation in terms of Jansen's (2017) categories (see Chapter 5 for details). One of these conceptions frames decolonisation as a way to de-centre Eurocentric/Western-centric knowledge in favour of marginalised knowledge systems (Sibanda, 2021, p. 184). Coupled to this view is the "additive-inclusive view" (Mohamed et al., 2020, p. 664), which stresses the importance of enhancing existing knowledge with alternative knowledge. Although this particular view is "soft" in the sense that it does not entirely negate Eurocentric and Western worldviews (Sibanda, 2021, p. 182), Mohamed et al. (2020, p. 664) point out that it is one preferred by critics of universal thinking. The third view entails "an engagement view" (Mohamed et al., 2020, p. 664) of science that resists the dismissal of disadvantaged or marginalised communities from participation in scientific projects.

We are of the view that regarding AI through a decolonial lens is also critical (Mohamed et al., 2020). In this respect, we too identify "sites of coloniality" (Mohamed et al., 2020, p. 666) that reflect problematic or contentious applications of AI. In view of their inevitable entanglements, the specific sites we highlight are not necessarily discussed as separate issues chapter by chapter. These sites include, but are not limited to, AI's diversity crisis and how this crisis impacts AI policies and strategies on the continent, AI ethics, human agency in the era of AI, the new world of work, education, the natural environment, as well as machine and deep learning which may reflect algorithmic biases. Given that AI forms part of a much larger socio-technical system that includes political, economic, and ecological dimensions, these problematic sites are also viewed through Smith and Neupane's (2018) typology of the (potential) risks of AI, which they have conceptualised specifically with developing countries in mind. The typology includes job losses (owing to automation), biases, loss of privacy (as a result of digital surveillance), and threats to human agency and thus to democracy (Smith and Neupane, 2018, pp. 11–12). These are similar to the "sites of coloniality" we consider as well as to the "areas of policy action" outlined by the United Nations Educational, Scientific and Cultural Organization/UNESCO (2020a) (see Chapter 3). To this typology, we add an additional risk which entails harm to the natural environment. We also consider what digital citizenship should encompass to address the challenges inherent in the various risks so that we are able to move closer to AI ecosystems in which principles such as justice, transparency, beneficence, and privacy are adequately addressed. This is an endeavour that dovetails not only with Metz's (2007) theoretical interpretation of an African ethic referred to in the previous section but also with responsible AI for Africa which encompasses "the ethical, transparent and accountable use of AI technologies in a manner consistent with user expectations, organisational values and societal laws and norms" (Delmolino and Whitehouse, 2018, p. 3).

Although we certainly acknowledge the importance of studying how the human body interacts with AI-enabled technologies, the focus of this book is on digitalisation as viewed through a socio-technical framework – on how the use of digital technology such as AI intersects with issues such as power, equality, and gender, among others. It therefore does not include discussions of socio-materialism – or how "[a]ssemblages of human flesh and nonhuman actors are constantly configured and reconfigured" (Lupton, 2017, p. 201). We recognise that socio-material and socio-technical systems appear to be synonymous terms, but Leonardi (2012) offers useful definitions of these terms to distinguish them: while socio-materiality describes the "[e]nactment of activities that meld materiality with institutions, norms, [and] discourses" (Leonardi, 2012, p. 42), for example, scholars who explore socio-technical systems do so with "[r]ecognition of a recursive [...] shaping of abstract social constructs and a technical infrastructure that includes technology and materiality" (Leonardi, 2012, p. 42).[3]

We turn now to a consideration of some of the underlying factors that have negatively impacted AI's diversity crisis and to how these factors to some degree stymie African nations' efforts to draft policies or strategies that ensure respect for diversity in AI ecosystems.

NOTES

1 This conception of decolonisation is not without its problems since it is regarded as a soft option in the realm of decolonised curricula (Jansen, 2017).
2 Note that our emphasis here is on individualism and not on individuality (cf. Comaroff and Comaroff, 2002).
3 Alistair Mutch offers some interesting insights into socio-materiality, critiquing it for "a failure to be specific about technology and a neglect of broader social structures" (2013, p. 28).

Chapter 2

Intricate intersections

AI's diversity and governance crises

INTRODUCTION

While holding numerous benefits for society, both the design and deployment of AI exhibit diversity problems, which are not only eroding human agency, identity, and privacy, to name just a few substantial concerns, but also threatening marginalised and vulnerable populations' already precarious socio-economic circumstances (cf. Lee, 2018, p. 252). Within a socio-technical framework, there is an urgent need to understand the complex intersections of these problems – particularly those related to race, gender, power, and labour – when it comes to what Gurumurthy and Chami (2019) aptly describe as "the wicked problem of AI governance" (Gurumurthy and Chami, 2019, p. iv), which among other things reflects the current exclusion of Global South participation in AI policy debates in which ethical issues predominate (cf. Gwagwa et al., 2020, p. 16). As considered in this chapter, AI's diversity crisis in Africa is to some degree rooted in colonial exploitation by the Western world and perpetuated by post-colonial political economies on the African continent that are disposed towards maintaining historical inequalities. We posit that decolonisation of policy discussions around ethical AI should be harnessed to counter ongoing inequalities and inequities within the AI ecosystem, where the former describes the uneven distribution of resources and the latter to biased or unfair practices. In the absence of Africa-inclusive AI policies, the continent runs the risk of being unable to address the power imbalances and value ladenness associated with AI-enabled technologies.

THE PORTRAYAL OF AI AS WHITE

Even a superficial Internet search of how AI is framed in the media, in popular culture, and in the technology industry reveals a propensity to depict this technology as white, and this is borne out by a number of scholars whose research interests run the gamut from applied ethics to engineering science (Bartneck et al., 2018; Cave and Dihal, 2020). Drawing on critical race

DOI: 10.1201/9781003276135-2

theory, which sheds light on covert racism, Cave and Dihal (2020) theorise that the stylisation of AI as white may in part be ascribed to the white racial frame which attributes intelligence, power, and professionalism to white culture. With respect to intelligence, if we review the history of Western thought in the context of Europe's colonisation of the territories in the seventeenth century, then intellectual justifications were generated for the colonial project. In this regard and in "The Problem with Intelligence", Cave (2020) recounts that in order to validate imperialism, Aristotle's argument that "the intellectually superior are by nature masters, and the intellectually inferior by nature slaves" (p. 34) was enthusiastically adopted by colonisers. Indeed, the supposed intellectual differences between the white race and so-called inferior races gained traction in the eighteenth and nineteenth centuries when scientists began experimenting with mental measurement, debating, among other issues, whether the taken-for-granted intellectual superiority of whites was cultural or biological (Cave, 2020, p. 30). By the twentieth century, psychologists such as Binet (1905), Terman (1916), and Brigham (1923) were developing intelligence quotient (IQ) tests, the origins of which can be traced back to eugenics, a pseudo-science which flourished at the beginning of the twentieth century and which posited that only superior races qualify for procreation, given their desired genetic traits (Cave, 2020, p. 31; cf. Levine, 2017, p. 1). Terman, a Stanford psychologist who did not attempt to camouflage his sexist and racist views, believed that IQ tests reflected the superiority of educated white men (Cave, 2020, p. 31), while Brigham's Scholastic Aptitude Test, designed to determine students' readiness for college in the United States, effectively gave only (wealthy) whites access to prestigious universities (cf. Neklason, 2019). Essentially, the use of IQ tests normalised the notion that white men reigned supreme owing to their superior mental aptitude.

Although the belief that race and intelligence are linked has been discredited (Harrison, 2020; Sternberg, Grigorenko and Kidd, 2005), scientific racism (or "race science") has not entirely disappeared and is reflected in a number of studies that Winston (2020, p. 427) observes evade discussing race by foregrounding "national IQ" instead (Lynn and Vanhanen, 2002; Rindermann, 2018). According to culture and media studies scholar Gavin Evans, who has written extensively on issues relating to race and IQ, *A Troublesome Inheritance* published by Nicholas Wade in 2014 "must rank as the most toxic book on race science to appear in the last twenty years" (Evans, 2018, para. 5). Wade (2014) argues that race is about biology, that the evolution of the human brain is different across races, and that different racial averages in IQ tests support these differences (cf. Absar, 2020; Evans, 2018). In a review of the book, Philip Cohen (2014) laments what he describes as the author's "mistreatment of Africa" (Cohen, 2014, p. 5). Claims such as African people's "violent, short-term, impulsive behavior typical of many hunter-gatherer and tribal societies" (Wade, 2014, p. 188) requires transformation and that Africa has failed to "develop the ingrained behaviors of trust, nonviolence and thrift that a productive economy requires" (Wade, 2014, p. 188) abound in the book.

Given the persistent racialised notion of an intelligence hierarchy, it is unsurprising that depictions of AI as white permeate the AI industry: mainstream white culture simply cannot imagine an intelligent machine that is not a white machine (Cave, 2020, p. 34; cf. Cave and Dihal, 2020, p. 13). (As an aside, mainstream white culture also cannot seem to envision a black machine that is not treacherous or aggressive. One fascinating study conducted by Bartneck et al. (2018) provides empirical evidence that AI-driven technology may be racialised. In their study, Bartneck and his colleagues adapted Correll et al.'s (2002) "shooter bias" paradigm to determine whether or not their research participants perceived the "agent" – a Nao robot whose colour was calibrated – as being racialised. As was the case in the original study, participants were tasked with shooting an armed "agent" but refraining from shooting him if he was unarmed. Bartneck et al. (2018) concluded that "participants were quicker to shoot an armed Black agent than an armed White agent, and simultaneously faster to refrain from shooting an unarmed White agent than an unarmed Black agent" (p. 201), thus confirming their initial conjecture that people may automatically categorise robots on the basis of race.)

To return to the white racial frame, mainstream white culture also perceives whites as the only race that has the capability and power to lay claim to professionalism, which may be defined as the ability to successfully carry out professional jobs in areas such as medicine, science, law, and academia (Cave and Dihal, 2020, p. 13). The stereotypical notion that whites possess "general mental capability" (Gottfredson, 1997, p. 13) to execute professional tasks extends to the AI sector too which, according to a sobering report on AI's diversity crisis, is currently dominated by white men who tend to be wealthy and technically qualified (West, Whittaker and Crawford, 2019, p. 6). That Western nations exercise control over the AI industry is evident in the lived experiences of AI experts such as Moustapha Cisse, head of the Google Artificial Intelligence Research Lab in Ghana and Nigerian scholar Tejumade Afonja, who studies the intersection of privacy and security in machine learning. BBC journalist Mary-Ann Russon (2019, para. 9) reports that in his blog, Moustapha Cisse narrates, "I have had papers accepted at meetings but been unable to attend because Western countries [...] denied me a visa". In addition, when Moustapha Cisse attended an AI conference in Spain in 2016, fewer than ten of the 5000 conference goers in attendance were black (Cisse, 2018, p. 461). Tejumade Afonja, along with other African AI experts, was denied a visa to attend a major conference on AI in Canada in 2019 (cf. Knight, 2019). Whatever the nature of these denials was, a senior writer for *Wired* observes that scenarios such as these point not only to the West's domination over AI-driven technologies but also to imbalances with regard to expertise and knowledge that "may end up continuing to skew the technology, biasing algorithms toward first-world perspectives" (Knight, 2019, para. 7).

WOMEN IN (EMERGING) TECHNOLOGY

In addition to exhibiting a race problem, AI, like other technologies, also has a gendered dimension, which should be considered in terms of "[the] fetishization of intelligence" (Cave, 2020, p. 3) during the colonial era as it relates to women and to post-colonial legacies that unfortunately continue to affect their advancement in emerging technologies such as AI. As noted above, one colonial era myth perpetuated by colonisers of African countries was that white men were mentally superior to black men, and this "gendered [idea] of an intelligence hierarchy" (Cave, 2020, p. 32) extended to women too as they were infantilised and their intellectual prowess negated (cf. Jaiyeola, 2020, p. 9). Throughout Africa, the stereotypical notion that women cannot perform successfully in STEM disciplines (science, technology, engineering, and mathematics) or in information and communications technologies (ICTs) means that gender-based digital exclusion remains endemic (cf. Borgonovi et al., 2018). Specific reasons why women in Africa continue to struggle to occupy careers in emerging technologies include insufficient access to ICTs and digital technologies as opposed to men (cf. Sicat et al., 2020), inadequate access to good education in STEM disciplines, and economic and socio-cultural obstacles (Gbedomon, 2016, p. 3). In addition, and as communicated in a 2020 African Academy of Sciences report, stereotyping as well as patriarchal and sexist attitudes further contribute towards gender disparities in STEM disciplines in Africa. Highly problematic is that

> masculine gender role stereotypes orient boys to acquire mastery, skills and competence, explore the physical world, figure out how things work, and are likely to be involved in activities that emphasize problem solving, status, and financial gain [...].
>
> (Mukhwana et al., 2020, p. 13)

Table 2.1 illustrates just some of the academic studies, industry reports, and media accounts generated over the past few years which draw attention to women's lack of access to the myriad opportunities afforded by the fourth industrial revolution (4IR) in Africa. In cases where we refer to "Sub-Saharan Africa" (SSA), we are aware that although this term is regarded by the United Nations as one that reflects African countries or territories that are located either fully or partially south of the Sahara, it is a designation that activists and academics working in the field of African-centred scholarship may perceive as derogatory and racist (Ekwe-Ekwe, 2012).

It would be misleading to suggest that the fourth industrial revolution constitutes an effective antidote to dismantling gender stereotypes in the (future) world of work – that the old order, as it were, will simply be replaced by post-gender structures (Bhatasara and Chirimambowa, 2018, p. 25). In fact, writing in the context of technology and tracking the history of each industrial revolution, Mudau and Mukonza (2021, p. 91) note that since

Table 2.1 Gender disparities in Africa in the context of the fourth industrial revolution

(Case) studies/media articles/ policy briefings/reports/surveys	*Findings*
Gbedomon (2016): a case study of how women may be empowered in technology in Kenya	"[D]espite an exponential growth in ICT penetration in Africa [...] there remains a critical ICT gender gap" (Gbedomon, 2016, p. 2).
Bhatasara and Chirimambowa (2018): a study of gender and the future of work in southern Africa	"[T]he fourth industrial revolution is most likely to perpetuate the structural nature of inequality" (Bhatasara and Chirimambowa, 2018, p. 25).
Letsebe (2018): an *ITWeb* media article on the representation of women in the ICT field in Africa, including South Africa	"According to the *2017 Global gender gap report*, only 13% of [South African] graduates in the STEM fields are women" (Letsebe, 2018, para. 4).
Adams (2019): an article published in *The Conversation* focusing on the risks women face in the fourth industrial revolution in Africa/South Africa	For women, the digital divide is particularly wide, as they do not enjoy high digital literacy levels and struggle to gain access to the Internet: "This suggests that women may be left out of increasingly digital work opportunities too" (Adams, 2019, para. 11).
Chiweshe (2019): a policy briefing on social inclusion and gender justice in the emerging fourth industrial revolution in Africa	"The 4IR thus does not provide any promise of improvement for women in low-skilled professions. These jobs are also at risk of automation [...] thus further increasing women's vulnerability" (Chiweshe, 2019, p. 5).
Dicks and Govender (2019): a study of feminist visions of the future of work in Africa	"[...] available statistics on the digital divide indicates that it is clearly gendered, and in some instances, is worsening" (Dicks and Govender, 2019, p. 4).
Majama (2019), organiser of the African School on Internet Governance (AfriSIG): a survey of the digital divide in Africa	"...Africa remains the only continent whose digital gender gap has widened since 2013" (Majama, 2019, para. 2).
Elhag and Abdelmawla (2020): a study of women's participation in science, technology and innovations (STIs) in Sudan	"[T]he participation of women in STI has remained low all over the world in general but particularly in [Sub]-Saharan Africa (SSA), where women are greatly underrepresented in [the] STI ecosystem" (Elhag and Abdelmawla, 2020, p. 97).
Jantijies (2019): a paper reviewing the role of tertiary institutions in increasing women's participation in STEM disciplines	"The lack of a diverse academic and research STEM workforce in educational institutions [...] leads to perceptions of STEM as a male-dominated domain. Academic institutions thus find it difficult to attract and retain women both as students and as employees in STEM" (Jantijies, 2019, p. 275).
Givá and Santos (2020): a study of women's participation in the STI ecosystem in Mozambique	"The gender-based STI ecosystem is weak in Mozambique, and policies and strategies are yet to be operationalized with appropriate implementation instruments" (Givá and Santos, 2020, p. 93).

(Continued)

Table 2.1 (Continued) Gender disparities in Africa in the context of the fourth industrial revolution

(Case) studies/media articles/ policy briefings/reports/surveys	*Findings*
Kisusu, Tongori and Madiany (2020): a case study of women's participation in the fourth industrial revolution in Tanzania	In the context of the fourth industrial revolution in Tanzania, "women participated more than men in agricultural production and unpaid domestic activities" (Kisusu, Tongori and Madiany, 2020, p. 13).
Libe (2020): a *CGTN* media article on promoting women's access to the technology industry in Africa	"In Sub-Saharan Africa [...] the overall female labor-force participation rate has reached 61 percent, yet women constitute only 30 percent of professionals in the tech industry" (Libe, 2020, para. 2).
Mulrean (2020): a study of women's roles in the fourth industrial revolution in Kenya, Nigeria, and South Africa	"Women [in Africa] run the risk of being left behind by advances in AI" (Mulrean, 2020, p. 19), representing a small percentage of those employed in the sector.
Rizk (2020): an article on AI and gender in African countries such as Egypt, Tunisia, Kenya, and South Africa	"While 60 percent of African women own a mobile telephone, only 18 percent have Internet access, and over 200 million remain unconnected" (Rizk, 2020, para. 4).
Sokona (2020): a gendered assessment of the STI ecosystem as it pertains to agricultural research and training in Mali	"All the indicators assessed exhibited large gender imbalance in favour of men" (Sokona, 2020, p. 63). (The indicators referred to are, among other things, human capital, and research and development in the STI ecosystem.)
Mudau and Mukonza (2021): a study of gender inequality in the area of ICTs	"[...] the gender inequalities in the technological discourse remain an old problem manifesting in the new context (4IR) and it is most likely to persist" (Mudau and Mukonza, 2021, p. 99).

gender disparities have not been resolved by each successive revolution (cf. Prisecaru, 2016), the fourth industrial revolution will in all likelihood maintain patriarchal structures too. Despite adopting an anti-colonial stance, post-colonial political economies in Africa continue to maintain historical gendered inequalities, given that men remain in control of resources that include land, capital, labour, and technology (Chimedza, 2019, p. 99).

Attempting to find technological solutions to gender disparities is unrealistic and utopian, as they simply cannot address power imbalances in society (Thakur, Brandusescu and Nwakanma, 2019, p. 333). Although, and as indicated in the previous chapter, *ubuntu* may constitute a useful framework within which to challenge gender inequalities in the deployment of AI (Mhlambi, 2020), feminist scholars and gender activists alike aver that this philosophy may be employed to further marginalise women, citing its deeply embedded patriarchal stance that in their view perpetuates the "oppression of African women" (Keevy, 2009, p. 41; cf. Kolawole, 2004). By contrast, scholars such as Chisale (2018, p. 4) are of the view that as a concept *ubuntu* is porous and thus open to misuse by individuals who may have personal reasons for establishing and maintaining gender inequalities. Still others perceive this philosophy as one that paradoxically "empowers women on the one hand by advocating for notions of equality and human dignity and, on the other hand, oppresses them by perpetuating [...] masculine authority and patriarchal values" (Chisale, 2018, p. 1). *Ubuntu* feminism is considered in the next chapter, but it is worth noting here that a number of scholars outside the field of AI argue that it could be employed to address concerns around gender disparities (Cornell and Van Marle, 2015) while also bridging the gap between indigenous knowledge systems and Euro-Western systems (Sefotho, 2021). Related to African feminism, *ubuntu* feminism as a theoretical lens reflects a number of principles that development studies scholar Gretchen Du Plessis (2019) argues include a "feminist ethics of care", "a mutually obligated life", "justice as equality", and "a call to social action" (Du Plessis, 2019, p. 43). We contend that these principles, while not reflecting an exhaustive list of the tenets of *ubuntu* feminism, nevertheless intersect both with ethical considerations around AI and with the development of policy frameworks around it.

ALGORITHMIC EXPLOITATION

The obstacles described so far reflect the coloniality of race and gender, since the effects of colonialism on black men and women persist in the present. Coloniality is founded on colonialism (Mohamed et al., 2020, p. 662) and may be defined as the continued practice of maintaining control over people, thereby depriving them of what is rightfully theirs (Mohamed et al., 2020, p. 662). If coloniality reproduces hierarchies of race and gender, then it is to be expected that it also replicates algorithmic exploitation on a number of levels.

Two manifestations of coloniality in the field of AI revolve around algorithmic oppression and algorithmic bias (Mohamed et al., 2020, p. 666), where the former entails one group of people being privileged over another through the deployment of algorithms and the latter to unrepresentative datasets that ultimately generate discriminatory outcomes. Numerous examples of the links between algorithmic oppression and colonial racism stem from North American scholarship (Hao, 2020, para. 7), but several instances have recently been published in African literature too. In an increasingly widely cited study entitled "Algorithmic Colonization of Africa", Abeba Birhane (2020a, 392) describes how Facebook (Meta) announced in 2016 that it was in the process of using AI and machine learning to map population density across much of Africa, the aim being to facilitate humanitarian agencies' efforts to locate households and distribute aid to them. Is such a venture entirely innocuous? Birhane (2020a) argues that although she is not averse to AI-enabled technology *per se*, what she rejects is Facebook's decision to assign itself as an authority to control and map the African continent: "Facebook [...] assumed authority over what is perceived as legitimate knowledge of the continent's population (Birhane, 2020a, p. 392).

What is problematic about the technology monopoly of companies such as Facebook and others (e.g., Google, Microsoft, and Amazon) is that they effectively reduce or block African countries' attempts to design and deploy their own AI technologies (Birhane, 2020a, p. 394). Africa needs to shift away from being a mere consumer of AI-driven technology manufactured by the West and East and produce AI tools that will address the continent's unique political and socio-economic challenges. The technology monopoly or "global tech oligarchy" as Benyera (2021, p. 77) refers to it should also concern us, given that many of the AI products produced and then utilised are not custom-made for African countries' profiles and specific contexts. One example of the importance of context emerges in the health sector. In one study, Owoyemi et al. (2020) found that although AI for healthcare in Africa may be beneficial, its implementation is not necessarily feasible, since this technology may be inaccessible owing to inadequate infrastructure such as lack of access to electricity to run the technology. What is more, the quality of healthcare tools based on AI depends on the quality of the training datasets employed, and this may be costly (Owoyemi et al., 2020, p. 2): "[w]ithout critical assessment of their relevance, the deployment of Western eHealth systems might pose more harm than benefit" (Birhane, 2020a, p. 396) to people living in Africa. In addition to algorithmic oppression, algorithmic bias refers to algorithms that are "built to automate procedures and [are] trained on data within a racially unjust society [...] replicating those racist outcomes in their results" (Hao, 2020, para. 7). According to a recent UNESCO report entitled, "I'd blush if I Could: Closing Gender Divides in Digital Skills through Education" (West, Kraut and Ei Chew, 2019), it is undeniable that AI (training) datasets also reflect gender biases, thus further exacerbating the stigmatisation and marginalisation of women.

A second manifestation of coloniality involves so-called clickworkers or ghost workers (Mohamed et al., 2020, p. 668), who are typically located in previously colonised countries and who carry out short-term labour to assist AI-driven technology in the so-called gig economy (cf. Anwar and Graham, 2020a, p. 1270; Hao, 2020, para. 8). The existence of ghost workers not only puts paid to the myth that AI does not necessitate human support but also "neatly extends the historical economic relationship between colonizer and colonized" (Hao, 2020, para. 8). To expand on this, while the term "gig economy" might be fairly new, as a phenomenon it is not, originating in "older, unregulated labour extraction methods" (Flanagan, 2017, p. 378), reflecting neoliberal economic models that have their roots in colonialism. Interrogating how women are precariously located within the gig economy in the context of southern Africa, Chimedza (2018) argues that not only does Africa's feminist movement have to contend with the hegemony of these neoliberal models, but it also has to confront the fact that some African states continue to adopt them: ghost workers who operate in countries such as those on the African continent include women who are generally poorly paid by technological and financial oligarchies and have dismal prospects for secure employment and career advancement (cf. Gray and Suri, 2019). Conceptualising the notions of "precarity" and "vulnerability" in the context of the gig economy in Africa, Anwar and Graham (2020a) report that as is the case across the globe, trade unions that could protect ghost workers from labour exploitation are virtually non-existent in Africa.

On the issue of labour, it is predicted that in light of quantum leaps in the development of machine learning and AI-enabled systems, automation will ultimately make significant inroads not only into manufacturing as is becoming manifestly evident across the globe but also into arenas that demand highly skilled training. At this stage, it is not possible to predict with any degree of certainty to what extent automation and intelligent automation (IA) will destabilise labour markets in Africa: they remain unknown quantities, while dire projections about the impact of automation have not necessarily been realised. In this regard, Smith and Neupane (2018, p. 79) correctly point to the threat of bank tellers being replaced with automated teller machines in the 1970s as one that did not ultimately materialise. Nevertheless, it is clear that unlike high-income countries, those in the Global South lack the resources and infrastructure necessary to absorb significant job losses through both automation and IA (Smith and Neupane, 2018, p. 79). Women are particularly vulnerable to these phenomena, as reported in a UNESCO report (2020b) on AI and gender equality. According to this report, research conducted by the International Monetary Fund and by the Institute for Women's Policy research signals that in the context of automation, it is women and not men who run the risk of being displaced in jobs that entail performing clerical and administrative tasks (UNESCO, 2020b, p. 4). Any discussion of labour markets and the future world of work for women on the African continent has to entail an acknowledgement of both historical and post-colonial imbalances: while the former resulted in disregarding African economies' informal sectors that were mainly populated by women, the latter has further aggravated women's marginalisation in these sectors. As economist and researcher Naome Chakanya (2018) puts it,

> Sadly, in the post-colonial period, most, if not all, of the African governments did not address the dual and enclave nature of their economies, but rather continued to marginalise and neglect the informal economy, further entrenching poverty and the feminization of the working poor. (p. 45)

Coloniality in the arena of AI is also patently evident through beta-testing (Mohamed et al., 2020, p. 668), a phenomenon that encompasses

AI systems being tested on vulnerable and marginalised groups to prove or invalidate their efficiency for users in high-income countries. A particularly insidious example of this kind of testing occurred when Cambridge Analytica tested its algorithms on the Nigerian and Kenyan elections in 2015 and 2017, respectively, before using them in the United States and the United Kingdom. The algorithmic tools deployed in Nigeria and Kenya not only resulted in election interference but also damaged social cohesion (Mohamed, Png and Isaac, 2020, p. 668; cf. Ekdale and Tully, 2019, p. 30). We argue that beta-testing of this nature also points to a related site of coloniality political scientist Matthew Crosston (2020, p. 149) refers to as "cyber colonisation", which he describes as a "dangerous fusion of artificial intelligence and authoritarian regimes" (Crosston, 2020, p. 149). Crosston (2020, p. 149) reports that under its Belt and Road Initiative (BRI), China has embarked on an aggressive marketing campaign of its AI-enabled technology in a number of countries that include North Africa and Sub-Saharan Africa and questions whether this campaign is an innocuous one or whether it in fact obfuscates more sinister strategic goals. Gravett (2020a) notes that "this first foray of Chinese AI technology into Africa is occurring free of the ethical and legal questions that are raised in more developed markets" (p. 162). What is problematic is that a number of countries that include Algeria, Botswana, the Côte d'Ivoire, Ghana, South Africa, Uganda, and Zimbabwe are themselves making extensive use of China's facial recognition software to control their citizens through covert surveillance of their movements (Jili, 2020, para. 2). Companies that manufacture and deploy this type of technology may do so ostensibly in the name of bringing about social good. Huawei, for example, sells its surveillance technology to African countries as part of its "Safe Cities" project (Jili, 2020, para 7). Enticements to purchase and deploy AI systems on the continent may also include promises in the form of extensive loans and assistance on technological or infrastructural levels (Gravett, 2020a, p. 162). What complicates deciding whether or not BRI initiatives in Africa are detrimental is that they continue to be upheld as an answer to the continent's need to address its major infrastructure deficits and digital connectivity issues. Yet this desperate need aimed at improving strategic economic goals obscures some of the detrimental consequences of the BRI, as addressed in Chapter 4.

ECOLOGY IN TWO SENSES OF THE WORD

Thus far, we have discussed imbalances on the continent as they relate to various social, economic, and political disparities but have yet to address the need for AI ecosystems that are also ecologically sensitive. Here, we are referring not to ecology in the material sense of the word but to ecology in its environmental sense – to the necessity of ensuring that neither the development nor deployment of AI damages natural environments that are already precariously positioned, given ongoing threats such as nutrient loading, climate change, pollution, overexploitation, and deforestation with its associated habitat loss. Under colonialism, African countries saw their natural resources mercilessly extracted and their land and animals exploited in the name of profit making and imperialism. In essence, nature was instrumentalised to serve the colonial project (Richardson, Mgbeoji and Botchway, 2006). In a post-colonial world, what we have inherited in some parts of Africa is "a continuation of inappropriate centralised government-decision making and frequent reliance on cumbersome, authoritarian modes of regulation, which together tend to disenfranchise local communities closest to nature" (Richardson et al., 2006, p. 416). The design and application of AI are to some degree an extension of colonialism, since many African nations are making use of Western technological giants' innovations that are not necessarily eco-friendly. For instance, African countries' appetite for AI products has exacerbated environmental risks in the form of raw material extraction for the production of lithium batteries and ever-higher carbon emissions owing to the consumption of huge amounts of energy. Although Research ICT Africa recently published a technical report by Sey et al. (2021) in response to a call by the International Development Research Centre (IDRC) to consider policies and capacities for responsible AI development in Sub-Saharan Africa, it does not address environmental concerns. Similarly, while the African Union's "Digital Transformation Strategy for Africa (2020–2030)" refers to the importance of investing in "green ICT" (African Union, 2020, p. 13) as well as to the "environmental benefits" (African Union, 2020, p. 35) that digital technology could help achieve, it does not indicate exactly how this should be operationalised. Given this neglect, we later consider how a relational ethics of care such as that encapsulated within the framework of indigenous environmental movements may be harnessed within AI ecosystems to further the well-being of both human and non-human entities in Africa.

ETHICAL AI AND POLICYMAKING

Another significant site of coloniality relates to the ongoing hegemony that developed countries exercise over regulatory policies governing AI: even a cursory online search of AI policies or strategies signals that most have originated in the Global North (Adam, 2019), thus excluding developing countries from adding their voices to what ethical AI should encompass. A systematic review of AI ethics guidelines by Jobin, Ienca and Vayena (2019) indicates that countries in the Global South such as Africa, South America, and Central America have been left out of the debate, which translates into an outright dismissal of indigenous knowledge systems and values (cf. Mohamed et al., 2020, p. 670). To put this reality into perspective, out of the 84 documents on the ethical management of AI collected by Jobin and her colleagues, more than a third originated from the United States and the United Kingdom, while countries such as Africa were substantially underrepresented (cf. Roche, Lewis and Wall, 2021, p. 644). In describing this state of affairs, Hogarth (2018, para. 1) goes so far as to refer to the existence of "AI nationalism", according to which AI superpowers are dominating policy frameworks and regulations pertaining to AI without taking African countries' social and economic realities into consideration (cf. Mohamed et al., 2020, p. 670). A concrete example of data imperialism is the European Union's passing of the GDPR – the General Data Protection Regulation – which aims to regulate how personal data is collected, stored, and utilised. According to Mannion (2020, p. 685), the European Union has signalled that if other countries that want to communicate with digital users in the Union do not comply with the GDPR's regulatory framework, their access to trade in that part of the world hangs in the balance. This type of data imperialism poses a conundrum for Africa: the domestic laws of countries on the continent simply do not accommodate the GDRP's demands, and "[e]ven if they did, several African countries lack stable judicial branches to enforce such laws, [while] many do not have the technological infrastructures or expertise to ensure ongoing compliance" (Mannion, 2020, p. 685).

To compound matters, the drafting of ethical standards relating to AI is in its infancy on the continent (Gwagwa et al., 2020; Nayebare, 2019; cf. Nayebare, 2019, p. 52), with Kiemde and Kora (2021) recounting that although 160 documents conceived around AI ethics have been submitted by various countries, Africa's contribution to this body of knowledge remains low. They also report that in terms of developing AI policies or strategies, a mere 17 countries on the continent have ratified legislation pertaining to data privacy and protection.

Of course, this does not mean that African countries are doing nothing to establish (in)formal norms, policies, or strategies to govern different aspects of AI technology. Despite the challenges surrounding AI governance, several countries on the continent have begun taking the necessary steps to harness its benefits and/or regulate its deployment. Some of these efforts across just a few countries are summarised in Table 2.2, providing a snapshot, as it were, of a variety of activities, albeit that many are not necessarily mandated at a national level.

A more recent initiative that is aimed at bringing together African partners with a view to leading the endeavour to initiate projects in the context of AI policymaking is Artificial Intelligence for Development Africa or AI4D Africa. This organisation's endeavours are considered in Chapter 7.

Some might argue that Africa's efforts to leverage the benefits of AI by adopting policies and strategies generated in the Global North constitute a chimera (Mudongo, 2020), given that many countries lack the technological infrastructure to accommodate this and other types of technologies. AI has been framed as two-edged (Besaw and Filitz, 2019) in the sense that while it

Table 2.2 Some African countries' norms, policies, or strategies around AI technologies

Country	*Norms/policies/strategies*
Botswana	Botswana has established a Science, Technology and Innovation Action Plan aimed at preparing for the fourth industrial revolution and fostering job creation for its citizens (Effoduh, 2020).
Cameroon	Cameroon launched its first AI training centre at the University of Yaounde in 2019 (Effoduh, 2020).
Egypt	Egypt's AI strategy revolves around a specialised AI faculty at Kafr-El-Sheikh University aimed at educating innovative AI thinkers to bolster the country's knowledge economy (Sawahel, 2019). Egypt has also established agreements with companies such as Microsoft and Teradata with a view to ensuring that AI applications are safe and sustainable (Effoduh, 2020).
Ghana	Ghana called for a ban on AI-driven lethal autonomous weapons systems (LAWS) in 2015, and many other African countries have supported this call, including Algeria, Egypt, Namibia, Sierra Leone, Uganda, and Zimbabwe (Stauffer, 2020).
Kenya	In 2018, the Kenyan government established a task force to create strategies around areas of AI related to cybersecurity, election processes, and service delivery, among other things (HolonIQ Report, 2020). Kenya has also launched an open-data portal that makes government datasets (on education, poverty, and sanitation, for example) freely available to its citizens.
Lesotho	The National University of Lesotho's Department of Mathematics and Computer Science offers research on AI projects as well as AI systems (Effoduh, 2020).
Mali	Water, Peace and Security (WPS) based in the Netherlands has designed AI-driven technology that is able to predict conflict hotspots that may cause water insecurity. Mali has partnered with WPS to make use of this technology (Burton, 2019).
Mauritius	Mauritius has a formal national AI strategy reflected in the Mauritius Artificial Intelligence Strategy, the Digital Government Transformation Strategy 2018–2022), as well as the Digital Mauritius 2030 Strategic Plan (Gwagwa et al., 2020).
Namibia	Namibia has formed partnerships with the United Nations Educational, Scientific and Cultural Organization/UNESCO and the Southern African Development Community to equip its citizens with the skills it will require to deal with the demands of the fourth industrial revolution (Effoduh, 2020).
Nigeria	Nigeria has established the National Agency for Research in Robotics and Artificial Intelligence (NARRAI), which aims to equip Nigerians with AI skills to grow the country's economy (Nayebare, 2019).
Rwanda	Rwanda Vision 2020 is a development programme established at government level that is aimed at improving women's involvement in STEM education (Jantijies, 2019).
Senegal	Following the example set by Tunisia, Senegal launched the Senegal Startup Act, which is in keeping with what is referred to as the Digital Senegal 2025 strategy. The Senegal Startup Act does not focus solely on AI but also fosters both innovation and entrepreneurship in the country (Jackson, 2019).

(*Continued*)

Table 2.2 (Continued) Some African countries' norms, policies, or strategies around AI technologies

Country	*Norms/policies/strategies*
South Africa	In 2017, South Africa's Department of Trade and Industry established The Future Industrial Production & Technologies (FIP&T) unit to assess the impact of the fourth industrial revolution and ensure that the government has the capacity to deal with the challenges of the revolution (Moitse et al., 2018). South Africa is working on developing strategies around data protection. One such strategy is reflected in The Protection of Personal Information Act 4 of 2013 (POPIA), which regulates the processing of personal information. South Africa has approved the Centre for AI Research (CAIR), which carries out research into different aspects of AI (Effoduh, 2020). The CAIR has nodes at North-West University, Stellenbosch University, the University of Cape Town, the University of KwaZulu-Natal, and the University of Pretoria.
Tunisia	In 2018, Tunisia established an AI Task Force and Steering Committee to promote an AI ecosystem in the country that will enable sustainable development and job creation (HolonIQ Report, 2020).
Uganda	Uganda has partnered with the United Nations to employ drones to evaluate the conditions of people fleeing from countries such as Nigeria and South Sudan (Effoduh, 2020, p. 5). Government health workers in Uganda make use of an SMS-based technology that enables them to keep track of and report on essential medicines, thus creating a link between hospitals and the drug chain (Kamulegeya et al., 2019).
Zambia	In Zambia, the government approved the country's so-called Smart Zambia e-Government Master Plan in 2018 that focuses, among other things, on developing ICT infrastructure and drafting effective legal and policy frameworks around ICT (Chilembo and Tembo, 2020, p. 35).

may be beneficial, one aspect of its "dark" side as already touched on relates to the continent being dictated to on a policy level. Mudongo (2020) cautions that a number of African nations have embraced the policy narrative of the World Economic Forum (WEF), a narrative that is wholly out of touch with each nation's unique challenges and values. In this regard, a number of scholars have raised concerns about the WEF's role, describing this body as "one of the world's most influential generators of neoliberal thought and policy" (Domingo, 2015, p. 100). Scholars also refer to guidelines for AI ethics that "are paternalistically positioned" (Adams, 2021, p. 184) as a universal solution to developing countries' challenges. The WEF's white paper on "AI Governance: A Holistic Approach to Implement Ethics into AI" published in 2019 certainly offers useful guidelines on creating ethical AI that is based on human-centric governance. Yet, a close reading of the paper also signals that matters relating to inequities, inequalities, and the future of work – which are just a few of Africa's most pressing problems – are not considered in any in-depth manner in the context of the Global South. In fact, a reference to "developing countries" appears only once in the 20-page document. Although the paper refers a number of times to the need to ensure that ethical AI embraces diverse values, cultures, and religions, for example, how this should be done in the context of the Global South is not explicitly addressed.

A HUMANISTIC VIEW OF ETHICAL AI GOVERNANCE

It should be clear that the overview provided above is by no means intended to convey the notion that a "one-size-fits-all" answer to generating policies around ethical AI in Africa should be adopted (cf. Kiemde and Kora, 2021, p. 2). After all, Africa reflects 54 sovereign nations that are pursuing their own individual AI goals aligned with their socio-economic agendas and unique cultures (cf. Nayebare, 2019). What we are proposing is that Africa's perennial inequalities and inequities cannot be solved by simply appropriating Euro-Western policies or strategies. These issues are also not amenable to technological solutions alone: AI's perceived risks for the continent demand ethical AI policies that underscore the human dimension far more than they do the technological element, a position shared by a number of scholars (Birhane, 2020a; Kiemde and Kora, 2021; Nayebare, 2019; Ndemo and Weis, 2017). This is a socio-technical view that acknowledges that AI technology is entrenched in complex social systems and that as agents, humans are both its designers and consumers (cf. Kroes et al., 2006, p. 806).

If the African continent is to harness the social and economic opportunities afforded by AI, then it is imperative for the continent to develop mature policies around this technology that will prevent the further entrenchment of both historical injustice and structural inequities. In order to foster such policies, we propose a decolonisation of AI that is not related to an "us versus them" (or "West and East versus the rest") mentality but rather to one that favours human-centred AI – "a truly global AI" (Mohamed, 2018, para. 21) as it were – and does not dismiss new voices: "this form of decolonisation asks us to work together to use existing knowledge, but to explicitly recognise the value of new and alternative approaches" (Mohamed, 2018, para. 9). Commensurate with this approach is that of web software engineer Michael Nayebare, who adopts Gertjan Van Stam's (2016) argument that what is required in Africa to reap the benefits of emerging technologies is "[a] 'we paradigm' that fosters community engagement and justice" (Nayebare, 2019, p. 50). Van Stam (2016) further asserts that this humanistic paradigm should reflect *ubuntu* – "communal love" (Nayebare, 2019, p. 52) or "humaneness" (Mnyaka and Motlhabi, 2009, p. 63) – so that humanity drives technology rather than the other way around. Several other scholars and technology experts have called for Africa to adopt this paradigm and include Mhlambi (2020), Langat, Mwakio and Ayuku (2020), and Maumela, Ṋelwamondo and Marwala (2020).

Although few researchers to date have examined aspects of AI through a socio-technical lens (Behymer and Flach, 2016; Borenstein, Herkert and

Miller, 2019), it is one that allows for the interrogation of how technical, social, and institutional dimensions are entangled and how they, in turn, intersect with ethical concerns. In this chapter, we have barely begun to scratch the surface of these complex entanglements, and so next examine them in more detail when we consider AI's ethical issues, taking into account AI's dual nature whereby it functions as a physical object that cannot be disengaged from human action (cf. Kroes et al., 2006, p. 806).

Chapter 3

AI ethics in and for Africa

Some preliminary thoughts

INTRODUCTION

Africa comprises 54 sovereign countries, nearly 1.4 billion people, 3000 indigenous groups, and approximately 2000 languages, making it more diverse than most continents (cf. Appiah, Arko-Achemfuor and Adeyeye, 2018; Kiemde and Kora, 2021). Given this diversity, it would be both presumptuous and impractical to propose a unified AI strategy that could address the risks and benefits AI holds for the continent. Even were Africa a more homogenous continent, imposing a standardised set of ethics guidelines on it would constitute devaluing the importance of pluralistic voices about AI in favour of universal, so-called objective knowledge about it that ignores "complex and changing interdependence and co-relations" (Birhane, 2021, p. 3). Globally, a flurry of attempts to formulate and impose abstract, generalisable AI principles on communities means that scant attention has been paid to AI's complex entanglements with developing countries' unique social, political, economic, ecological, and ethical landscapes. In the context of gender, for example, a number of UNESCO Dialogue participants – such as Rediet Abebe, Genevieve Macfarlane Smith, and Daniela Braga – have noted that the normative principles generated thus far fail to adequately address gender inequalities and inequities (UNESCO, 2020b, p. 12). What is essential is for African nations to pay careful attention to how ethics guidelines are crafted to address issues of risk such as digital surveillance, algorithmic bias, and job losses through automation (Wairegi, Omino and Rutenberg, 2021). Given the proliferation of ethics guidelines in recent years, countries run the risk of simply picking and choosing those principles that they believe – or choose to believe – suit their (unscrupulous) agendas without interrogating whether these principles do in fact address serious socio-economic realities. Floridi (2019, p. 186) calls this "ethics shopping". It is a practice according to which "private and public actors [...] shop for the kind of ethics that is best retrofitted to justify their current behaviours, rather than revising their behaviours to make them consistent with a socially accepted ethical framework" (Floridi, 2019, p. 186).

DOI: 10.1201/9781003276135-3

Building on the last section of the previous chapter, we briefly consider disconnects between the content of AI guidelines or debates and what the current realities are. We then review both major and minor theories emanating from philosophical traditions that are at present being employed to guide the drafting of ethical principles before considering what ethics guidelines for Africa could encompass.

DISCONNECTS

The ethical principles around AI that have been generated through numerous policy debates and proposals emanate from medical ones: while this is fairly logical, given that the principles inherent in both bioethics and digital ethics demand acknowledgement of "new forms of agents [...] and environments" (Floridi and Cowls, 2019, p. 5), they do not take cognisance of some of the fundamental differences between a medical environment and the AI ecosystem. We add that a significant piece of the principle puzzle that is missing is clearer recognition of AI's dual nature – of its (potentially exploitative) contexts of use and its architecture, which is a moving target. This vacuum may seem surprising but is one noted by Hagendorff (2020, pp. 103–104), who has observed the propensity in ethics guidelines to regard AI technologies as separate from the social, institutional, and ecological settings in which they are embedded. This tendency may in part be due to the prominent place that a utilitarian approach to AI ethics has enjoyed until now, with its emphasis on technology as an agent of change. Such an emphasis aligns with technological determinism (Greene, Hoffmann and Stark, 2019, p. 2122), a reductionist theory that perceives technology as defining culture and society (Dusek, 2006, p. 84). The avoidance of framing AI as an isolated artefact is critical, given that transhumanist techno-orientated principles are currently challenging what it means to be human (Lamola, 2021).

Exacerbating the consequences of treating AI as being disconnected from its contexts of use is the rise of so-called ethics washing (cf. Wagner, 2018), whereby institutions pay mere lip-service to guidelines by simply substituting a code of ethics for government regulation. Such a façade obstructs serious debates around AI and renders ethical codes ineffectual. We have already seen by way of just a few examples thus far how the absence of regulation of AI manifests itself on the African continent in the form of algorithmic bias, algorithmic oppression (Birhane, 2020a), and cyber colonisation (Crosston, 2020).

Thomas Metzinger, a theoretical philosopher at the University of Mainz who, as a member of the High Level Expert Group on Artificial Intelligence (HLEGAI), helped formulate the EU Commission's "Ethics Guidelines for Trustworthy AI" (2019), laments a significant omission reflected in the document, which is the small number of ethicists invited to contribute to its drafting. In a news article published in *Der Tagesspiegel* on 8 April 2019, Metzinger noted the document's preponderance of voices from industry, calling the guidelines "lukewarm, short-sighted and deliberately vague" before stating that "They ignore long-term risks" and "gloss over difficult problems ('explainability') with rhetoric" (Metzinger, 2019, para. 5). The same criticisms could be levelled against many of the AI documents currently circulating around the globe (cf. Hagendorff, 2020, pp. 99–100).

Coupled to these disconnects is what was established in the previous chapter, namely, that the ways in which ethical AI has been framed in policy discussions and draft documents present a dilemma for countries in Africa, given that Western-centric framing of various issues has little, if anything, to do with these countries' unique contexts as they relate to societal, cultural, political, economic, and ethical dimensions. That current policies or strategies are not Africa-inclusive is a significant oversight, yet few scholars in Africa have begun considering what such policies or strategies should encompass (Gwagwa et al., 2020; Kiemde and Kora, 2021; Langat, Mwakio and Ayuku, 2020; Mhlambi, 2020; Nayebare, 2019; Ormond, 2019).

ETHICAL DEBATES AROUND AI

At present, Western conceptions of utilitarianism and deontology constitute the two approaches that are commonly employed in discussions around ethics and AI, with the former approach – simplistically put – "weighing the consequences of actions" (Kazim and Koshiyama, 2021, pp. 3–4) and the latter asking whether or not these actions follow strict moral imperatives. Utilitarian approaches are "based on the idea that one can, at least in theory add up the aggregate utility and disutility resulting from a particular course of action" (Stahl, 2021, p. 20), whereas deontological ethics underscores the *motivation* behind an action (Stahl, 2021, p. 20). A third, less dominant approach is virtue ethics (sometimes referred to as teleological ethics), which subordinates the consequences of an agent's actions to his/her virtues. Working within a socio-technical framework, Vallor (2016, p. 48) refers to these virtues as "technomoral practices" that include but are not limited to honesty, self-control, justice, empathy, care, and perspective (Vallor, 2016, p. 120). Of importance is that virtue ethics focuses not so much on technology *per se* as it does on the traits, dispositions, and choices of its architects (Hagendorff, 2020; Maiyane, 2019); it may guide developers in "[framing] ethical discussions in the context of changing social conditions (possibly brought on by AI technologies)" (Yu et al., 2018, p. 5531). In practice, and in the context of AI-enabled robotics, individuals who employ utilitarianism would consider a robot's actions or behaviours in terms of the outcomes of these actions or behaviours, whereas those using a deontological framework would ignore the consequences and underscore the robot's intentions instead (Pagallo, 2016, p. 210). Individuals who view AI through the lens of virtue ethics would be more concerned with whether or not the designers of the robot considered moral values such as self-control and empathy than with programming it with moral reasoning. If a more technical approach is taken, then from a utilitarian point of view, emphasis would be placed on tenets such as autonomy and beneficence, while a deontological lens would accentuate "what is obligatory, permissible, or forbidden" (Pagallo, 2016, p. 210).

Utilitarianism remains a prominent framework within which to interrogate ethics and AI, given that the tenets underlying it such as accountability or privacy "are most easily operationalized mathematically" (cf. Hagendorff, 2020, p. 103) and thus oriented towards technical solutions.

Metz (2021, p. 59) notes that in terms of morality, what makes utilitarianism attractive is that it calls for maximising overall good or well-being, and therefore "appears to capture the logic of many everyday decisions that at least 'modern' western people make". However, some scholars view utilitarianism as being at odds with the African philosophical tradition (Mhlambi, 2020). As is the case within a utilitarian framework, from an African perspective, it is crucial to improve the well-being of others. However, if improvement is achieved at the expense of others, it compromises human dignity: "[i]mproving others' well-being does have a place in African thought, but most often insofar as doing so can sometimes be a way of expressing respect for people who have a dignity" (Metz, 2021, p. 61). Loss of dignity typically occurs when, for example, AI's so-called goodness benefits certain sectors of society but perpetuates systemic inequality in others (cf. Mhlambi, 2020, p. 22). A concrete example is the use of AI in the medical field: it is by now well documented that AI for healthcare developed in high-income countries is beneficial to people in those countries because it has been customised to accommodate their medical needs and socio-economic landscapes, but that its deployment in developing countries is either impractical or prejudicial (Alami et al., 2020). With respect to the former, low- and middle-income countries in Africa might lack sufficient expertise and infrastructure in the form of Internet penetration or energy grids to manage the implementation of medical AI, particularly in under-resourced rural areas (Guo and Li, 2018). With respect to the latter, training datasets in AI models might not represent diverse, heterogenous population groups, thus aggravating biases and disparities in healthcare (Kaushal, Altman and Langlotz, 2020). Working in the areas of computer ethics, computer science, and artificial intelligence, Burton et al. (2017) echo concerns about utilitarianism's "goodness" and lack of sensitivity to different contexts, arguing that

> These shortcomings limit our ability to have substantive ethical discussions, even insofar as everyone assents to utilitarianism; a shared reliance on the principle of "the greatest good for the greatest number" does not help us agree about what goodness is [...].
>
> (Burton et al., 2017, p. 27)

In contrast to utilitarianism, deontology underscores strict principles and rules by questioning how they are employed in decision-making processes and whether or not they help realise specific social goals (Burton et al., 2017, p. 26). It is thus a legalistic approach that essentially represents "an end of ethics, not ethics itself" (Resseguier and Rodrigues, 2020, p. 3), culminating as it does "in different principles, norms or requirements" (Resseguier and Rodrigues, 2020, p. 3). While useful, such an approach remains divorced from what Hagendorff (2020, p. 114) refers to as "a situation-sensitive ethical approach based on virtues and personality dispositions, knowledge expansions, responsible autonomy and freedom of action" (cf. Martinho, Herber, Kroesen and Chorus, 2021). It is for this reason that a number of researchers working in AI contexts have begun calling for a greater – though not necessarily exclusive – emphasis on virtue ethics (Bynum, 2006; Hagendorff, 2020; Maiyane, 2019; Neubert and Montañez, 2020; Stahl, 2021), which has three varieties referred to as neo-Aristotelian virtue ethics, agent-based ethics, and care ethics, although the last variety is not regarded by all scholars as an example of virtue theory, given that it does not focus on dispositions but on relations (Held, 2006, p. 19).

Neo-Aristotelian virtue ethics stresses contemporary issues (Constantinescu et al., 2021) and is based on Aristotle's virtue ethics according to which an agent needs virtues (such as honesty and generosity), intellect, and external goods (such as health and friends) to achieve *eudaimonia* or human flourishing (Sakellariouv, 2015). By contrast, agent-based virtue ethics derived from the work of Michael Slote (2021) shifts the focus to an agent's inner world in the sense that the moral status of an action performed by the agent is subject to the agent's character traits and motives (Slote, 2021). Care ethics is perceived as a variety of virtue ethics that places emphasis on a caring agent concerned with establishing and preserving relationships with others through moral emotions. As noted, it is for this reason that the ethics of care is perceived by some scholars as one that cannot be subsumed under virtue theory: "[v]irtue ethics focuses especially on the states of character of individuals, whereas the ethics of care concerns itself especially with caring *relations*" (Held, 2006, p. 19).

Agent-based virtue ethics is not considered in any detail here, not only because a significant objection to its application in the field of AI pertains to its inability to tell an agent exactly what to do (cf. Sakellariouv, 2015), but more importantly because at this stage AI has no moral intelligence (Roff, 2019). In this regard, if intelligence is a nebulous concept, then artificial general intelligence (AGI) is more so. In *The Myth of Artificial Intelligence* published by Harvard University Press, Erik Larson offers some intriguing insights into why AGI may never come to fruition, arguing that "As we successfully apply simpler, narrow versions of intelligence that benefit from faster computers and lots of data, we are not making incremental progress, but rather picking the low-hanging fruit" (Larson, 2021, p. 2) as it were. Although AI is revolutionising myriad sectors of society, the notion of a super-intelligent machine is a pipe dream, with current algorithms reflecting "the intelligence of an abacus: that is, zero" (Floridi, 2018, p. 157). Contributing to an in-depth report on AI, computational scientist Peter John Bentley (2018) dispels a number of myths that AGI poses an existential threat to our existence as do many other scholars and industry experts (Casilli, 2021; Natale and Ballatore, 2020; Shevlin et al., 2019). In terms of moral reasoning, several researchers are exploring AI's potential to exhibit moral intelligence (Arnold, Kasenberg and Scheutz, 2017; Bringsjord, 2013), but while sophisticated machine learning models produce any number of outcomes, moral reasoning is currently not one of them (cf. Brożek and Janik, 2019; Haidt, 2003). Researchers tend to refer, not to moral principles for AI systems but to value alignment which entails determining "how to ensure that AI systems are properly aligned with human values and how to guarantee that AI technology remains properly amenable to human control" (Gabriel and Ghazavi, 2021, p. 1). What we take from studies of AGI and moral reasoning is that it would be a pointless exercise to draft ethics guidelines around AI that has not yet materialised – or, as Barbat (2018, p. 4) fittingly expresses it, "to give in to the temptation to legislate on non-existent problems".

Turning to the next variety of virtue ethics, part of neo-Aristotelian virtue ethics' appeal hinges on its emphasis on *eudaimonia*, which shifts our focus "from act-based moral philosophizing to the indispensability of moral character traits or virtues, happiness, or human flourishing" (Bozkaya, 2021, p. 289). It is also attractive because unlike traditional philosophies which speak mainly to human agency, it has the capacity to accommodate non-human agents such as AI-enabled social companions and digital assistants (cf. Bynum, 2006, p. 158). Interestingly, despite neo-Aristotelian virtue ethics' emphasis on eudaimonic well-being, some scholars contend that this philosophy also accommodates non-human entities such as animals (Dzwonkowska, 2018; Sandler, 2007, 2018, p. 18). Dzwonkowska (2018) avers that while anthropocentric concerns around virtue ethics imply a dismissal of non-human animals and that biocentrism and ecocentrism have not resolved these concerns, ecological issues could be addressed through ethical anthropocentrism, as opposed to ontological anthropocentrism and epistemological anthropocentrism. Dzwonkowska (2018) further recommends a focus on Ronald Sandler's (2007, 2018) ethical approach to the environment which is non-anthropocentric and underscores both ecological sensitivity and a flourishing of humans. Delving at length into discussions of environmental philosophy and of differences between anthropocentric and non-anthropocentric perspectives are beyond the scope of this book, but Cafaro (2015) and others (e.g., Dursun and Mankolli, 2021; Karlsson and Edvardsson Björnberg, 2021) offer some useful insights into such discussions. As noted elsewhere, the humanistic approach we ascribe to is one that eschews AI-centrism and anthropocentrism in favour of an approach that acknowledges care for humans and the natural environment. As will be noted a little later, such an approach is to some degree consistent with the tenets of a relational ethics of care, rather than with environmental virtue ethics.

If eudaimonist virtue ethics is compared to African conceptions of ethics such as *ubuntu*, then both acknowledge self-realisation (Metz, 2012, p. 100). However, a fundamental difference is that while virtue ethics foregrounds individualist virtues, African theories place the emphasis on communitarianism (Metz, 2012, p. 99). Expressed slightly differently, virtue ethics stresses self-sufficiency, or how the individual can better him- or herself, whereas African thought is more concerned with how that betterment helps the community flourish too (Ogunyemi, 2020, p. 3). A distinction can thus be made between self-regarding and other-regarding virtues. Since virtue ethics accentuates the former, some might argue that it to all intents and purposes also stresses individualistic self-interest, although this is open to debate. In the context of responsible innovation (RI), for example, Steen, Sand and Van De Poel (2021) propose a contemporary neo-Aristotelean virtue ethics according to which individuals involved in RI are required to find solutions to society's problems and develop their well-being. They argue that the virtues extolled by Aristotle, though agent-centred, implied benefit to both individuals and the communities in which they lived. In this sense, virtues "are meant to contribute to living together in a *polis* – in a society; we are *zoon politikon*, social animals" (Steen et al., 2021, p. 7). This aligns with Annas' (1993) claim that "an ethics of virtue is at most formally self-centered or egoistic; its content can be as fully other-regarding as that of other systems of ethics" (Annas, 1993, p. 127).

A RELATIONAL ETHICS OF CARE

Without delving into complex debates as to whether or not virtue ethics and an ethics of care are at odds with one another, debates that would fill volumes[1], we posit that although virtue ethics may include an other-regarding dimension, it nevertheless foregrounds individual self-sufficiency. This emphasis does not reflect an approach that speaks *directly* to how wider contexts, relationship networks, and dependencies intersect with AI systems on the African continent. As noted in the previous chapter, Africa exhibits significant disparities in terms of race, gender, power, and labour, for example, and so what is required is a focus not so much on technology as on one that takes into account how technology both benefits populations and puts vulnerable communities at risk. Similar to an African ethical approach such as *ubuntu*, we propose that debates and guidelines around AI and ethics could benefit from insights gleaned from a relational ethics such as an ethics of care, a proposal put forth by a number of scholars working in AI contexts (e.g., Asaro, 2019; Birhane, 2021; Cohn, 2020; Yew, 2021).

Conceived by Carol Gilligan (1982) in "response to morality, justice and judgement-centric ethical systems" (Cohn, 2020, p. 3), an ethics of care acknowledges not only the existence of intricate, potentially fraught entanglements between people and institutions but also the need for contemplative decisions and actions that take into account marginalised and vulnerable populations' well-being (Cohn, 2020). Having said this, we acknowledge a legitimate charge against an ethics of care approach, which is that it could be perceived as belittling and patronising. However, at any given time, we are all – to a greater or lesser extent – vulnerable to the perils of emerging technologies, and so the ethics of care we have in mind is one that recognises all of humanity within an AI ecosystem. Borrowing to some degree from Herring's (2014, 2017) argument in regard to the charge of condescension, we do not distinguish between "a carer and cared for" (Herring, 2017, p. 160). Instead, our conception of an ethics of care is one that looks to tenets that, unlike those reflected within utilitarian or Aristotelean frameworks, do not primarily accommodate the well-being and beliefs of "the most advantaged over everyone else" (Cohn, 2020, p. 3). We concede that an additional criticism of an ethics of care is that, having exhibited an initial focus on "family ethics" (Held, 2006, p. 18), it is still perceived as an approach that has bearing only on personal contexts, and yet recent formulations have seen it transcend private spaces to also accommodate social, political, and economic spheres. Hamington and Sander-Staudt (2011) have considered how care ethics may be employed in business, for example, while others have applied it in the context of human resource development (Armitage, 2018), political theory (Engster and Hamington, 2015), and more recently AI (Bhirhane, 2021).

What would an ethics of care in the context of AI look like? In Chapter 2, we proposed that *ubuntu* feminism could offer a useful starting point to begin answering this question, although it should be noted that we are not suggesting that feminism and African moral thinking can be conflated (cf. Hall, Du Toit and Louw, 2013, p. 29). We are also not proposing that a feminist ethics of care is necessarily superior to any other approach to ethical AI. As indicated, developmental studies scholar Gretchen Du Plessis (2019) describes *ubuntu* feminism as a theoretical framework that acknowledges, among other things, the importance of an ethics of care, "justice as equality" (Du Plessis, 2019, p. 43), the "mutually obligated nature of human existence" (Du Plessis, 2019, p. 44), and calls to social action. In Gilligan's (1982) framework, justice encompasses principles such as equality and rights, whereas care reflects equity and responsibility, among other things. Du Plessis' (2019) emphasis on justice as equality in our view aligns with Held's (1995) argument that while justice is critical, it should be applied within the framework of an ethics of care, which focuses on "contextual detail, on the self [...] in relation to others" (Clement, 1996, p. 110). How do justice and care conceptualised in this way apply particularly to the *deployment* of AI? How do mutual responsibilities and a call to social action apply? A few practical scenarios are presented here in an attempt to answer these questions and reflect sites that reproduce the type of coloniality discussed in previous chapters.

In the case of algorithmic biases related to race and gender, for example, one seemingly straightforward and innocuous solution is to simply "debiase" flawed and unrepresentative datasets, but there are limits to such an approach (Balayn and Gürses, 2021). For one thing, it constitutes a technocratic one that assumes all problems may be resolved through technical solutions. For another, it obscures power asymmetries as well as historical and post-colonial imbalances that often lie behind skewed or biased algorithms: it is certainly possible that discrimination is the result of algorithmic biases, but most harms "relate to how AI systems operate in a broader context of structural discrimination, recreating and amplifying existing patterns of discrimination" (Chander, 2021, p. 9). To compound matters, a technocratic approach means that whichever technology companies monopolise the markets and debiase their products in essence become "arbiters" of inequalities (Balayn and Gürses, 2021, p. 118). Challenging such a reality calls for resisting hierarchical power imbalances by asking "What is the product or tool being used for? Who benefits? Who is harmed?" (Birhane, 2021, p. 2). Working within a relational framework that embraces both relations and dependencies, Birhane (2021) offers some practical guidelines around non-technocratic solutions that include recognising that, as opposed to the most privileged, "minoritized populations (1) experience harm disproportionally and (2) are better suited to recognise harm due to their epistemic privilege" (Birhane, 2021, p. 5). Such an approach allows for the foregrounding of the redressing of algorithmic bias by the very population groups harmed by it (Birhane, 2021, p. 5). Social action could also take the form of calling for ethics guidelines that ensure that AI systems such as those in healthcare or in finance are initially tested for transparency, biases, and errors in a variety of contexts before deployment (Smith and Neupane, 2018, p. 15). In terms of the technical dimension, the design of AI models should be a participatory venture in that diverse communities comprised of the disproportionally impacted, civil society, and academia, among others, must have a role in developing them in the first place (Smith and Neupane, 2018, p. 15). All the strategies described here underscore the fact that AI algorithms do not represent a kind of universal, objective truth about the world we live in and that they therefore cannot be perceived as offering solutions to Africa's pressing challenges if deployed in isolation. The relational dimension of the ethics of care proposed here is one that recognises that sensitivity to diverse contexts and concrete action on the part of "society-in-the-loop" (Rahwan, 2018, p. 5) are essential to ensure the well-being of all who inhabit the AI ecosystem.

We argue that emphasis should be placed on all *AI-stakeholders-in-the-loop* to accent the importance of a plurality of voices and values within the AI ecosystem. Myriad expert, public, and private stakeholders such as those in government, business, civil society, and academia are either directly/indirectly involved in the design/deployment of AI systems or affected by these systems in implicit or explicit ways. Within the framework of an ethics of care, pluralistic perspectives of AI are, as Tschaepe (2021) puts it, critical to avoid "singular and obstinate [perspectives], in considerations of design, use, and ethical consideration" (Tschaepe, 2021, p. 106). One significant criticism of a pluralistic perspective might be that in terms of epistemology, it conveniently escapes the need to hold people to account when some or other action or behaviour is detrimental (Cooper, 2020). However, it is a perspective that reflects "a valuing of others and how they see the world and a desire to engage with others in respectful and empowering ways" (Cooper, 2020, para. 4). As noted in an earlier chapter, it is often disparate views rather than uniform ones voiced within AI contexts that may be helpful in exposing inadequacies in the debate around AI and ethics (Rudschies et al., 2020).

To briefly return to algorithmic bias, in their review of AI ethics guidelines, Jobin, Ienca and Vayena (2019) have noted that across all guidelines, commonly cited tenets include non-maleficence, transparency, justice, and privacy, to name a few (Jobin et al., 2019). This is to be expected, given that they originate from African and Western bioethics. Yet what Floridi and Cowls (2019) describe as "the crucial missing piece of the AI ethics jigsaw" (Floridi and Cowls, 2019, p. 8) is "explicability" or "explainability", which they argue needs to be conceptualised in two ways. On the level of ethics, explainability refers to "accountability", which constitutes an answer to the question "Who is responsible for the way [AI] works?" (Floridi and Cowls, 2019, p. 8). Epistemologically, it should be understood as "intelligibility", which explains how AI operates (Floridi and Cowls, 2019, p. 8). On an epistemological level, we need to be concerned that AI's modality is such that some machine learning systems are able to make independent "decisions" once their designers have specified their parameters and predominant goal (Yeung, 2019, p. 20). A significant problem inherent in this type of modality is the so-called black box conundrum reflected in artificial neural networks or ANNs, which renders their "decision-making" processes opaque (Carman and Rosman, 2021, p. 112): "[i]nstead of storing what they have learned in a neat block of digital memory, [ANNs] diffuse the information in a way that is exceedingly difficult to decipher" (Castelvecchi, 2016, p. 21). Black boxes are unfortunately susceptible to risks, which may include biases, whether deliberate or unintentional on the part of their designers. Several researchers have exposed instances of algorithmic bias towards Africans. In a study that focused on measuring racial bias in deep facial recognition, a research team conducted experiments on a dataset referred to as "Racial Faces in-the-Wild" and determined that specific algorithms reflected racial bias, with error recognition on African faces being twice that of error recognition on Caucasian faces (Wang et al., 2019). One of the conclusions the researchers reached was that racial bias was caused both by the training data and by aspects of the algorithms. In the context of "Gender Shades", a project aimed at assessing the accuracy of gender classification technology powered by AI, Ghanaian-American computer scientist Joy Buolamwini exposed this technology as generating both gender and skin-type biases.[2] One of the discoveries made by Reidsma (2016) when he conducted a study of algorithms in the context of students' use of library discovery systems was that biased results about race were generated when searches were made for "X in South Africa". In 2018, South Africans browsing the Internet became aware that searching for "squatter camps in South Africa" on Google yielded mainly images of white squatter camps, which does not reflect the reality that 1% of whites are poverty-stricken (South African Human Rights Commission, 2017/2018). The inaccurate portrayal of the country's socio-economic landscape was blamed on algorithmic bias, although of interest is that Jansen Van Vuuren and Celik (2019) argue that it is the data rather than the algorithm that could have been biased.

Given algorithmic inscrutability and biases, we propose that it is worthwhile paying far greater attention to applying explainability in the design and deployment of AI in Africa (cf. Carman and Rosman, 2021). This principle is one that "*prima facie* at least, is not based strongly on Western values" (Carman and Rosman, 2021, p. 111), given that it is not geared towards extreme individualism, which as we have already noted is anathema to the philosophy of *ubuntu*. Instead, because it partly emphasises *who* is responsible for how AI works, it puts the human back into AI, so to speak. Expressed a little differently, "explainability in AI is as much of a Human–Computer Interaction [...] problem as it is an AI problem, if not more" (Ehsan and Riedl, 2020, p. 450). Explainability also makes transparent what kind of data was collected as well as how and why it was collected in the first place.

With respect to AI's gender problem as it pertains to women in the (future) world of work and in educational settings, a technocratic approach to mitigating disparities would be to invest a great deal of money in AI-enabled technologies and infrastructure. The paradox, however, is that technological and infrastructural investment may exacerbate the digital divide which, as Lutz (2019) points out, exhibits more than one dimension or level. The first-level digital divide is one that we are all most familiar with, which is the absence of, or unequal access to, technologies. Of significance is that the legacy of colonialism is such that while Africa has invested in digital technologies like AI, progress has been frustratingly slow (McDonald, 2019). Further,

> Digital technologies [...] have been reflective of colonial tendencies through the fact that Western states, in an attempt to 'help' countries across Africa, have steered them towards the adoption of digital technologies and encouraged a dependency on them despite only a handful of benefits.
>
> (McDonald, 2019, para. 4)

Conceived of by Hargittai (2002), the second-level divide describes inequalities in terms of skills and uses that are mainly the result of socio-economic imbalances. Within such a divide, an important point to be made is that even if the technological infrastructure exists, the question remains "what's the point of infrastructure without the skills to use it?" (Kassongo, Tucker and Pather, 2018, p. 2). If access to skills is indeed available, are these skills accessible to all? Across Africa, systemic inequalities and inequities hamper the majority of women's attempts to develop AI and other digital skills (cf. Chiweshe, 2019, pp. 1–2), which are still acquired mainly by white males (cf. Smith and Neupane, 2018, p. 57). Both lack of access to AI-enabled technologies and the skills needed to use them hinder or entirely obstruct women and other marginalised people's opportunities to either enter the job market or gain access to STEM education. A UNESCO report on AI and gender equality published in 2020 and alluded to in the previous chapter is one of the few documents in circulation that offers an Africa-inclusive view of AI and ethics (cf. Gwagwa et al., 2020, p. 16). While the entire report is not explicitly couched within a relational ethics of care paradigm, it nevertheless outlines a number of key strategies or tenets that align fairly well with such a paradigm given their relational emphasis. A section of the report devoted to gender equality and AI principles specifically calls for "Feminist Internet Principles" (UNESCO, 2020b, p. 11) that, although focused on the Internet, could very well apply to AI contexts and are consistent with Du Plessis' (2019) conceptualisation of *ubuntu* feminism. Among other things, these principles advocate (1) recognising that technological spaces, like others, perform and impose patriarchal governance that must be resisted as well as addressed through, for instance, "democratizing policy making" and "diffusing ownership of and power in global and local networks" (UNESCO, 2020b, p. 11); (2) challenging as well as dismantling the neoliberal undertones evident in the technology sector through, for example, "cooperation, solidarity, environmental sustainability, and openness" (UNESCO, 2020b, p. 11); and (3) "[claiming] the power of technology to amplify women's narratives and lived experiences" (UNESCO, 2020b, p. 11). Importantly, and in the context of acknowledging diverse settings as they relate to language or abilities, for example, the report also calls for (4) "the right to code, design, adapt and sustainably use and reclaim" (UNESCO, 2020b, p. 11) technology as a response to sexism and discrimination. These principles clearly stress relationality, as opposed to rationality, and constitute a clarion call to justice within the framework of an ethics of care.

Notwithstanding the difficult task of attempting to find definitive evidence that powerful governments or other elites in a given country are violating data privacy or deploying AI systems to surveil its citizens and possibly clamp down on their movements, Hadzi and Roio (2019) aver that such crimes reflect both authoritarian and utilitarian tendencies on the part of "power elites" (Hadzi and Roio, 2019, p. 2). Although a recent phenomenon, AI surveillance in Africa is steadily increasing, with some African countries making use of facial recognition systems under the guise of establishing safe city projects (Mudongo, 2021, p. 1). What is problematic about such surveillance is that it is not undertaken in any transparent manner, while the risks it holds such as those related to data privacy issues and discrimination have not been adequately addressed in AI guidelines for Africa. Mudongo (2021) reports that while many countries in West and North Africa such as Ghana and Nigeria have risk mitigation frameworks in place to safeguard their citizens' safety and privacy as do several in south-central or southern Africa such as Botswana and Zambia, some countries in East Africa, including Uganda and Kenya, have not implemented any data laws to protect their citizens. In the context of humanised AI systems, Hadzi and Roio (2019) propose that within the framework of an ethics of care, restorative justice could be used as a means to address AI crimes. They argue that "Restorative justice might support victims of AI crimes better than the punitive legal system, as it allows for the sufferers of AI crimes to be heard in a personalised way" (Hadzi and Roio, 2019, p. 17). In a study focusing on China's AI programme in Africa, South African legal scholar William Gravett advocates a number of measures that need to be put in place to safeguard individuals from exploitative technologies. Among other things, "Lawmakers, civil-society leaders and technologists should press for appropriate safeguards to deal with the practical human-rights challenges arising from major AI-related programmes" (Gravett, 2020a, p. 168). Gravett (2020a, p. 165) points out that although Africans also have recourse to the African Union's Convention on Cyber Security and Personal Data Protection in the event of data crimes, it has been signed by only a handful of countries to date. By our last count, only 14 countries on the continent had signed the policy by May 2020.[3]

Returning to UNESCO's (2020b) document on AI and gender equality, a number of principles that once again fall under the banner of "Feminist Internet Principles" have been proposed to address issues related to data privacy/surveillance, data security and safety, and technology-related violence. These principles, which may be generalised to many vulnerable groups such as persons with disabilities, migrants, and refugees, reflect the need to attend to agency, memory, anonymity, youth/children, and violence in digital spaces. With regard to (5) agency, there is an urgent need to "build an ethics and politics of consent into the culture, design, policies and terms of technology" (UNESCO, 2020b, p. 14). Here, agency refers to the control that vulnerable and marginalised groups should exercise with regard to whether or not, and to what extent, they are willing to disclose the private and/or public dimensions of their lives. It also reflects "a rejection of practices by states and private companies to use data for profit and manipulate behavior, as well as surveillance practices" (UNESCO, 2020b, p. 14). Closely aligned with agency is (6) memory, which calls for people to have the freedom to manage all aspects of their digital information (from choosing who has access and who does not to deciding to delete their digital footprints) and (7) anonymity, which advocates the right for digital users to protect their identities (UNESCO, 2020b, p. 14). With respect to (8) children and youth, the UNESCO (2020b) document recognises the need for minors and young adults to have their voices heard in the contexts of safety, privacy, and access to positive (and we would like to add, accurate) digital data. The document also calls for (9) social action on the part of various stakeholders in the context of "technology-related violence" (UNESCO, 2020b, p. 14).

As noted in previous chapters, ethics guidelines have given little consideration to protecting non-human animals within the AI ecosystem (cf. Coeckelbergh, 2021; Owe and Baum, 2021), although there are some exceptions to this inattention. The *Montréal Declaration for the Responsible Development of Artificial Intelligence* (2018, p. 6), for instance, notes that in terms of the principle of well-being, "The development and use of artificial intelligence systems (AIS) must permit the growth of the well-being of all sentient beings", while UNESCO's (2020a) *Outcome Document: First Draft of the Recommendation on the Ethics of Artificial Intelligence* refers to the importance of both "animal welfare" and "the environment and ecosystems" (UNESCO, 2020a, p. 1). Under its *Recommendation of the Council on Artificial Intelligence*, the Organisation for Economic Co-operation and Development (OECD), endorses "responsible stewardship of trustworthy AI in pursuit of beneficial outcomes for people and the planet, such as [...] protecting natural environments" (OECD, 2019, p. 11). In the context of ethics guidelines for Africa, we have to date not detected any (draft) documents that pay explicit and detailed attention to AI and its impact on the natural environment. Although Smith and Neupane's (2018) white paper on AI and human development refers to Africa's challenges with regard to animal extinction, crop disease monitoring, climate change, waste, and flood impact, among others, emphasis is placed on how AI systems may be harnessed to address these issues to the benefit of humankind, rather than on how AI's design and application may be harming the natural environment. A few researchers on the continent have drawn attention to animal welfare in the context of AI. Nandutu, Atemkeng and Okouma (2021), for example, have noted that "the existing literature weakly focuses on [...] AI Ethics in wildlife conservation while at the same time ignores AI Ethics integration in AI systems for wildlife conservation" (Nandutu et al., 2021, p. 1). There is no doubt that although AI may be leveraged to protect the natural environment, its use requires high energy output (Coeckelbergh, 2021), while its consumption by humans is resulting in the extraction and depletion of the earth's natural resources (Khakurel et al., 2018).

To ensure ecological sensitivity in the design and implementation of AI technologies, we suggest that tenets be generated around an ethics of care that has not been acknowledged with any serious attention, in all likelihood because it is an approach embedded in "anticolonial ethics and epistemologies" (Whyte and Cuomo, 2016, p. 235). Environmental care ethics is a counter-response to legalistic and abstract thinking around the environment and stresses the notion that care for fellow humans and care for the natural environment go hand in hand. Given African countries' vast natural wealth and the threats to this wealth that AI's development and implementation pose, different notions of an ethics of care could be considered in the context of ethical guidelines, with one increasingly prevalent conception revolving around indigenous environmental endeavours. This type of environmental ethics of care stresses the importance of tapping into local and indigenous knowledge about ecological issues and thus embraces Mohamed et al.'s (2020, p. 664) call for a re-centring of knowledge systems that were previously devalued and negated under colonialism. Whyte and Cuomo (2016) point out that "The moral theories explicit in indigenous environmental movements [...] furnish guidance for decision making on action and policy in relation to environmental issues" (Whyte and Cuomo, 2016, p. 240). Within AI ecosystems, such guidance should encourage preservation of the natural environment through technological solutions that are guided by the local community (Sheppard et al., 2020). One such a solution "requires a shift in emphasis from outsiders' concerns to local concerns and knowledge" (Sheppard et al., 2020, para. 12). This shift aligns with indigenous environmental movements' contention "that intimate relationships of interdependence yield complex forms of moral and scientific knowledge" (Whyte and Cuomo, 2016, p. 240). In the context of AI in Africa, several projects around biodiversity and sustainability have been designed and deployed within local communities according to their values, beliefs, and knowledge systems (cf. Sheppard et al., 2020). Founded by Katy Payne, the Elephant Listening Project (ELP) deployed in central Africa is one that in our view embraces an indigenous environmental approach, given that it combines AI-enabled technology with traditional knowledge to combat illegal poaching of forest elephants in central Africa. Research in Africa that integrates AI technologies and indigenous knowledge systems is that conducted by Masinde (2015) and Akanbi and Masinde (2018), which focuses on the development and deployment of drought early warning systems (DEWS) in Kenya, Mozambique, and South Africa. In discussing the importance of local and indigenous knowledge, Masinde observes that as a response to scientific knowledge alone, "Contextualised innovations built by, with and for local people, have a higher chance of succeeding and Indigenous Knowledge Systems bridges this gap because it supports ways that are culturally appropriate and locally relevant to them".[4] The way in which this observation is framed – stressing the need for technological solutions to be developed by local communities – is significant, given that it alludes to the fact that participation by disenfranchised communities is insufficient as they also need to be empowered in the design of technological innovations. Transcending participation to embrace indigenous-built innovations is a way to redress past power asymmetries and create "a culturally-tailored, culturally-enriched and trustworthy environment for participation" (Peters et al., 2018, p. 100).

FEMINIST AND ENGAGED COMMITMENTS TO AI

The principles outlined here acknowledge complex co-relationships within the framework of an ethics of care, endorsing a humanistic view of AI that challenges, addresses, and redresses this technology's multiple impacts on humans, humanity, and the natural environment. They are ones that highlight the need to frame AI policy not in terms of technological hype, but within a framework that acknowledges that AI technologies must emanate from co-creation within given societies (Ulnicane et al., 2021, p. 160). They are also in our view consistent with Shaowen Bardzell's (2010) identification of specific qualities that she argues are critical in the context of human-computer interaction. These are "pluralism, participation, advocacy, ecology, embodiment, and self-disclosure" (Bardzell, 2010, p. 1305), which we propose collectively allow for what De Jaegher (2021) refers to as "an engaged epistemology" of knowing (De Jaegher, 2021, p. 849). We have already noted that within AI ecosystems, the design and application of AI-driven technologies require *pluralism* – the myriad and inevitably divergent perspectives of AI-stakeholders-in-the-loop from all sectors of society that resist the notion of a universal view of the AI ecosystem. Western-centric conceptions of AI should concern us: as Bardzell (2010, p. 1305) avers in the context of cross-cultural design, adopting a universal perspective carries with it the risk of negating communities' unique cultural and social attributes. With regard to *participation*, the principles we highlight reflect participatory endeavours that go beyond peripheral or passive involvement and call for stakeholders-in-the loop to actively partake in the design and deployment of AI. This means a shift away from the exclusionary notion in cognitive science that highlights "discrete, rational knowing to the detriment of engaged, human knowing" (De Jaegher, 2021, p. 847). Our call for diverse AI-stakeholders-in-the-loop to both resist and reform AI inequities and inequalities reflects *advocacy*, which entails questioning ethical issues in the development and application of AI that impact both its creators and consumers. The latter encompasses not only human users but also non-human entities in the AI ecosystem, and so we advocate for ethical tenets that include awareness of *ecology* in both its environmental and material senses. The principles we have underscored reflect *embodied* ethics, given that they acknowledge the need to consider AI's complex and often contentious intersections with a variety of disparities related, among other things, to gender, race, and the (future) world of work. This is consistent with feminist scholars' appeal for a relational ethics that does not decontextualise technology at the expense of people's lived experiences (cf. UNESCO, 2020b, p. 11). Finally, championing the necessity to make transparent how and why AI works as it does and to insist on accountability in its deployment aligns with Bardzell's (2010, p. 1307) quality of *self-disclosure*.

Several researchers, notably Floridi (2019) and Hagendorff (2020), have recently begun questioning *how* the principles reflected in ethics guidelines should be operationalised. This call to action is legitimate, given the rapid pace at which AI is developing across the world. However, we question if countries in Africa are quite ready for this shift, arguing that it would be premature and foolhardy to simply pick and choose tenets that appear to (conveniently) suit different needs and agendas. Considering that (1) current global guidelines are exclusive of African contexts and that (2) the fourth industrial revolution is playing itself out differently on the continent in the context of (3) major inequalities and inequities that may only be exacerbated rather than resolved through AI, there is an urgent need to avoid ethics shopping in favour of an approach that prudently and methodically embraces the inclusion of all.

Since AI is embedded in society, it is critically important not only to consider public perceptions of this technology but also to interrogate the various ways in which it is framed and implemented. With respect to the latter, stakeholders that include government, industry, and business have a penchant for describing, and – where they have the power or authority – for implementing AI in ways that tend to camouflage its ethical and epistemological impacts. Considering misleading or false perceptions and framings of AI is also vital, given that they "can lead to regulatory activity with potentially serious repercussions in society" (Cukurova, Luckin and Kent, 2020, p. 207). Perceptions, framings, and deployments of AI are considered next and juxtaposed with the social, economic, and political realities in many countries on the African continent. Recognising the disjuncts between what AI promises to offer and what is currently happening on the continent is critical and a reminder that, while crucial, generating AI ethics is not a panacea to various challenges on the continent (cf. McLennan et al., 2020).

NOTES

1 In this regard, see Sander-Staudt (2006), Halwani (2003), and Thomas (2011), for example. Some scholars call for care ethics to be subsumed under virtue ethics, while others argue that they should be entirely separate disciplines.
2 http://gendershades.org/overview.html
3 https://au.int/en/treaties/african-union-convention-cyber-security-and-personal-data-protection
4 https://www.cut.ac.za/news/prof-muthoni-masinde-presents-to-scifest-afri

Chapter 4

(Mis)perceptions of AI in Africa

Metaphors, myths, and realities

INTRODUCTION

Thus far, we have considered AI's intersections with Africa's complex socio-economic realities as well as interrogated how and why current ethics guidelines and debates are out of touch with these realities. We have also suggested partially addressing the continent's AI challenges through a relational ethics of care that recognises the importance of qualities such as advocacy, inclusivity, participation, and plurality within AI ecosystems. Foregrounding AI as a socio-technical phenomenon does not in any way imply a dismissal of AI as either a scientific or technical field of enquiry. However, the way in which the fourth industrial revolution is emerging in Africa is clearly dissimilar to how it is evolving in the Global North (cf. Wairegi, Omino and Rutenberg, 2021, p. 2) and subsequently demands an emphasis on AI for social improvement, rather than on AI for technological innovation alone (Kiemde and Kora, 2021, p. 5): in a recent exhaustive and sobering assessment of how the fourth industrial revolution is developing in Africa, political scientist Everisto Benyera offers the caveat that it "will not end or lessen [the continent's] challenges which are a product of centuries of being on the darker side of Euro-North American modernity" (Benyera, 2021, p. xi). To compound matters, in proposing a way forward for Africa to harness the benefits of AI, Gwagwa et al. (2021) report that Africa scores low in terms of AI readiness owing to the absence of a vision for AI deployment (Gwagwa et al., 2021, p. 5). Gadzala (2018, p. 1) goes so far as to declare that for most African countries, employing AI remains an *ignis fatuus*, given the absence in these countries of definitive reforms as they pertain to critical issues such as data privacy, governance, and digital education.

In the previous chapter, we referred to the importance of AI-stakeholders-in-the loop. In identifying these actors or stakeholders, Wairegi, Omino and Rutenberg (2021, p. 6) indicate that while primary (or expert) stakeholders such as AI developers and companies have "direct input" in the design and application of their products, secondary and tertiary stakeholders do not have this privilege. Secondary stakeholders include consumer groups and advertising companies that may enjoy some level of participation, while

DOI: 10.1201/9781003276135-4

tertiary actors made up of the general public in all probability experience little, if any, direct involvement in AI ecosystems. All stakeholders have intrinsic value (Wairegi et al., 2021, p. 5), but the focus of this chapter falls mainly on tertiary actors, particularly on excluded or vulnerable individuals who are most at risk of being harmed through the development and deployment of AI, although secondary actors are also considered, since they too often exhibit a layperson's knowledge of AI. The questions are *How do secondary and tertiary actors in Africa perceive AI in the first place?* and *Why does this matter?* While it is easy to answer the second question, it is far more difficult to address the first, given the dearth of sources that have attempted to explicitly gauge public opinion of AI on the continent. What complicates the issue is that defining AI is not a simple task and may be clouded by the hype-filled message that it constitutes a cure-all for human problems. Obscuring the realities of what AI is and what it does is the proliferation of misleading metaphors employed to describe its myriad aspects. Hype around AI and the various imaginaries employed to frame it neither comport with this technology's realities nor acknowledge Africa's diverse, social, political, and economic landscapes.

AFRICAN PERCEPTIONS OF AI

In recent years, the few papers or surveys that have to some degree either directly or indirectly addressed public sentiment towards AI in Africa include Neudert, Knuutila and Howard's (2020) large-scale survey of global perceptions of this technology's risks, Gwagwa et al.'s (2021) summary of the state of AI in Sub-Saharan Africa, and a 2020 poll of South Africans' perceptions of AI conducted by Viacom Global Insights South Africa. In their working paper, Neudert et al. (2020) stress that across the globe, societies are fairly ambivalent about AI's development, generally perceiving it in terms of risks rather than benefits, although they also note some key differences between regions. In contrast to other countries, people inhabiting North America and Western Europe, for example, are of the view that AI and robotics will bring about more harm than good. In terms of perceptions of threats associated with automated decision-making, people who live in Africa and East Asia, for instance, are less concerned about its pitfalls than those who occupy Europe or Latin America. Although they do not explicitly address public opinion, Gwagwa et al. (2021) note that according to a 2021 UNESCO survey of AI capacity building needs authored by Sibal, Neupane and Orlic, a mere 21 African countries have prioritised AI in their plans at national level (p. 6). Such a lukewarm response does not bode well for marginalised populations, who rely on the government to exploit AI's benefits while also mitigating its risks. Sibal et al. (2021, p. 66) cite fears around AI that include issues related to privacy, lack of human agency, bias, and gender discrimination. A 2020 poll conducted by Viacom Global Insights South Africa reports that although individuals aged 18–49 are equivocal about AI's capabilities, uncertain as to how the technology will directly affect them as well as concerned about possible job insecurity and forfeiture of human agency, most expressed the opinion that AI will ultimately be more beneficial than detrimental. In this regard, they cite some of AI's main advantages as enhancement of productivity on the part of humans and the eradication of faults as some of AI's main advantages (Dias, 2020). While useful, this poll targeted the online community, thus excluding at least 36% of South Africans who do not have access to the Internet (Kemp, 2020). Other Africa-centred industry reports briefly refer to people's fears around AI as they pertain to income inequalities, uneven development, and the entrenchment of injustice (Butcher, Wilson-Strydom and Baijnath, 2021; Pillay, 2020; Schoeman et al., 2021).

An informative study that explicitly addresses ordinary citizens' perceptions of AI in a framing theory approach is one by Donald Malanga (2019), who examined how Malawian women from academia, government, civil society, and the private sector conceptualise AI and human rights. Overall, perceptions of AI were mixed. Respondents were of the view that AI could improve their social and economic circumstances on condition that it was implemented according to ethical guidelines that stressed the protection of human rights. However, given that Malawi's constitution pays lip-service to gender equality in light of the perpetuation of patriarchal views (Malanga, 2019, p. 172), it is unsurprising that the respondents also identified specific fears around AI's deployment. These included concerns about facial or speech recognition software being exploited to curtail freedom of expression and encroach upon their privacy and about AI being used to discriminate against women based on race, gender, and ethnicity, effectively impeding them from gaining access to jobs or places at institutions of higher learning. Of significance is that Malanga (2019) not only reports that the respondents exhibited a basic knowledge of what AI entails but also notes low levels of awareness among them of what kinds of AI projects are currently being deployed in the country. What is more, none of the women reported participating in any of Malawi's AI initiatives, which revolve around machine learning, biometrics, and drones, to name a few (Malanga, 2019, pp. 174–175).

The sprinkling of academic studies or media reports that call attention to African perceptions of AI are those in the realm of the professions. This is particularly evident in the field of radiography, although it must be acknowledged that deployment of AI systems in healthcare on the continent at this stage pivot around pilot projects that are limited in nature (Owoyemi et al., 2020, p. 1). In studying African radiographers' perceptions of the use of AI in medical imaging practice, Antwi and his team found that while radiographers were generally positive about AI's application, noting improvements in diagnosis, patient safety, and research, among other things, they were fairly concerned about a number of issues. These included lack of proper training and data protection, fears around job loss, and concerns about radiographers' core skills and roles vanishing (Antwi, Akudjedu and Botwe, 2021, p. 1). In the field of academic librarianship, Lund et al., (2020) measured academic library employees' perceptions of the use of AI across six continents including Africa. They concluded that academic librarians – particularly those described as early adopters who tend to enthusiastically espouse innovations after careful deliberation – were overwhelmingly positive about AI's integration in the day-to-day operations of libraries. This finding is in marked contrast to one by scholars in Nigeria who found that while librarians acknowledged that AI could enhance their productivity and increase satisfaction among library users, they were concerned about being replaced by machines (Abayomi et al., 2021). With regard to the use of AI in the legal profession, Adeyoju (2018) notes that while references to AI may engender fear among lawyers, it is nevertheless being embraced – albeit on a small scale – by law firms in Kenya, Nigeria, South Africa, Tanzania, and Uganda to carry out mundane tasks related to legal research, processing of historical data, analyses of statistical information, and the like. In the field of journalism, perceptions of AI's role in journalistic practices are somewhat mixed. Allen Munoriyarwa and his colleagues point to "a deep-seated skepticism with AI in South African newsrooms" (Munoriyarwa, Chiumbu and Motsaathebe, 2021, pp. 15–16), which is partially the result of fears relating to ethical issues and job insecurity, while the media themselves refer to loss of decision-making on the part of editors and journalists as a significant concern (*Daily Maverick*, 28 September 2021). Some media outlets display positive sentiment, citing AI as a useful gatekeeper for moderating online news consumers' (vitriolic) comments (*News24*, 15 July 2020). Others offer a tongue-in-cheek view of AI, with Jay Caboz from *Business Insider SA* allaying journalists' fears of job insecurity by describing the paper's use of an AI writing assistant as a failed experiment, given that what it generated was "vague and flowery" as well as factually inaccurate (*Business Insider SA*, 17 October 2021). In the accounting profession, Kamau and Ilamoya (2021, p. 6) aver that while AI is a disruptive technology in Kenya, it should not pose a threat to basic tenets inherent in accounting, which is a view shared by the South African Institute of Professional Accountants (SAIPA) (Moyo, 2019), although Professor Rashied Small, executive for education and training at SAIPA, warns that accountants will have to complete technology-orientated training to keep pace with the profession's rapid digitisation (Small, 2019). Adopting a gender lens to interrogate the impact of AI systems on financial technology ecosystems (FTs) in Ghana, Kenya, Nigeria, and South Africa, Ahmed (2021) notes that "gender workforce disparity [...] contributes to gender barriers in the FTs that impact women's access to many financial services and resources" (p. 15). Ahmed (2021, p. 15) further predicts that AI's applications in such a setting will not remove biases related to gender and race.

TECHNO-UTOPIA VERSUS TECHNO-DYSTOPIA

It appears then that as is the case in other parts of the world, perceptions of AI on the continent are uncertain and are often framed as "techno-utopian or dystopian imaginaries" (Lupton, 2021, p. 1). The former imaginary is often ubiquitous, given that it fosters neoliberalism (Jones and Hafner, 2021), which not only emphasises a global order but also "confidently identifies itself with the future" (Brown, 2006, p. 699). Within such a context, the AI economy is characteristically framed as normative – "naturalised as *the* common-sense way of life [...] and a 'public good'" (Bourne, 2019, p. 113). With its emphasis on economic globalisation, this type of economy is one that typically ignores economic diversity and consequently exacerbates existing inequalities, subordinating social justice issues such as income gaps, gender inequalities, racial discrimination, and healthcare to the maximisation of profits (cf. George, 2020; Gurumurthy and Chami, 2019). As we have reiterated a number of times, only a few countries in Africa boast the requisite data ecosystems, infrastructure, and resources to tap into the AI economy and improve the socio-economic conditions of their citizens. Yet even in these cases, African digital giants are habitually state-owned, effectively excluding smaller players from entering lucrative markets (cf. Candelon, Bedraoui and Maher, 2021) and in some instances adding to the precarity and vulnerability of gig workers' lives. In terms of both scenarios, the existence of monopolies in the telecommunications industry offers a case in point. With respect to the former, "New entrants into some telecoms markets in Africa often complain of anticompetitive or unfair monopolistic behaviours adopted by dominant incumbent players" (Amadasun et al., 2021, p. 5). With respect to the latter, M-Pesa, for example, is a mobile money service provider in Kenya structured around thousands of agents that locals significantly refer to as "human ATMS" (McBride and Liyala, 2021, p. 3). This appellation underscores the reality that although M-Pesa has increased employment opportunities, supposedly allowing for leapfrogging over old technologies and affording financial inclusion, its informal workers' knowledge and behaviours are essentially traded for profit. Park (2020) contends that "narratives of technological leapfrogging and 'financial inclusion' elide the reliance of the digital economy on older formations of precarious, often feminized, reserves of labour power [...] whose progenitors can be easily discarded in the interest of profitability" (p. 917). Tinashe Chimedza, associate director at the Institute of Public Affairs in Zimbabwe (IPAZIM), identifies ECOCASH in this country and SAFARICOM in Tanzania as other local digital platforms that are worsening unjust gender relations in Africa: "when the bottom lines from the gig economy are announced from MPESA in Kenya, ECOCASH in Zimbabwe and SAFARICOM in Tanzania, the question that faces us is: Where are the women when the dividend cheques are written?" (Chimedza, 2018, p. 100).

It is certainly the case that African countries are seeking to exploit emerging technologies, striving to move beyond colonial and post-colonial abuses. Indeed, digital entrepreneurship on the continent is flourishing as a direct result of the fourth industrial revolution, and success cases abound. In this respect, and using a multi-site case study approach, Friederici, Wahome and Graham (2020) offer numerous examples of the boom currently taking place, to a greater or lesser extent, in 11 cities across Anglophone, Francophone, and Lusophone Africa. These cities are Abidjan, Accra, Addis Ababa, Dakar, Johannesburg/Pretoria, Kampala, Kigali, Lagos, Maputo, Nairobi, and Yaoundé. However, Friederici et al. (2020) also note that entrepreneurial ventures across the continent do not necessarily translate into breaking free from Africa's socio-economic legacies. This is in part because much of the digital infrastructure is in the hands of non-African companies (Hruby, 2018). While facilitating entrepreneurship on the continent, transnational digital platforms such as those in North America have nevertheless "strategically monopolized precisely the most scalable digital product categories, outcompeting upstarts from other locations based on financial advantages, multipronged scaling economies, and lock-in effects" (Friederici, Wahome and Graham, 2020, p. 218). Currently, American digital companies such as Amazon, Apple (Alphabet), Facebook, and Google dominate the continent, and it appears that Asian companies that include Alibaba, Baidu, Huawei, and Tencent are not far behind (Hruby, 2018). In fact, in marked contrast to European companies, US- and Asian-based companies enjoy the monopoly over data reserves across the globe, making it impossible for African countries to join the AI race at this stage. Technological ecosystems managed by monopolies and tethered to neoliberal economies embedded in already precarious social, economic, and political landscapes simply cannot and do not reflect inclusivity, participation, and plurality. In the context of the fourth industrial revolution, scenarios such as these make the continent highly susceptible, among other things, to what Benyera (2021, p. 15) refers to as "the (re)colonisation of Africa and Africans" (Benyera, 2021, p. 15).

In an earlier chapter, mention was made of China's Belt and Road Initiative (BRI), an enterprise ostensibly established to facilitate infrastructure projects in Africa and Europe. Given a number of significant deficits in Africa related to infrastructure and funding, the BRI appears to be a logical and attractive solution to many countries' strategic development goals. Among a host of benefits to Africa, the BRI "could contribute in meeting the continent's huge infrastructure financing requirement, estimated at US$130–170 billion per year" (African Development Bank [AfDB], 2018; Lisinge, 2020, p. 426).

To date, 42 African countries have agreements with the BRI (Shukra, Zhou and Wang, 2021), and specific developments such as Nigeria's 2,600-megawatt Mambilla Hydropower Station; railroad construction in Gabon, Mauritania, and Nigeria (Risberg, 2019, p. 45); and the Standard Gauge Railway initiative in Kenya (Wilson-Andoh, 2022) appear to showcase the advantages of these agreements. Although most BRI projects are aimed at energy and transportation, some focus on the ICT sector. At the opening ceremony of The Belt and Road Forum for International Cooperation five years ago, President Xi of China called for BRI initiatives that also facilitate innovation in the sense that they "pursue innovation-driven development and intensify cooperation in frontier areas such as digital economy, artificial intelligence, nanotechnology and quantum computing" (Xi, 2017, p. 5). In 2015, China launched the digital development component of the BRI, which it expedited once COVID-19 broke out. Among other things, this component involved the use of Chinese AI-enabled technology to power diagnostic systems and ensure that medical provisions were tracked (Lo, 2021).

While clearly beneficial to society, such digital pursuits in Africa have a shadow side. In this regard, one of the more treacherous forms of technological dystopia that we have touched on constitutes AI-enabled technologies that are slowly yet steadily being co-opted by digital dictatorships, which describe governments tracking, controlling, and undermining their citizens' freedoms (Gopaldas, 2019, p. 4). In a policy brief on digital authoritarianism, Polyakova and Meserole (2019) note that technology manufactured by China's ZTE (Zhongxing Telecommunications Equipment) is deployed in Ethiopia to surveil both journalists and the opposition, while in southern African countries such as Angola and Zimbabwe, Chinese technology such as that produced by CloudWalk is employed to monitor political opponents on a mass scale. According to a global survey of social media manipulation conducted in 2020, at least 81 countries make use of a specific type of trolling activity in the form of so-called cyber troops, 13 of which are African countries (Bradshaw, Bailey and Howard, 2021, pp. 1–2). Cyber troops, defined as "government or political party actors tasked with manipulating public opinion online" (Bradshaw et al., 2021, p. 1), work through social media platforms such as Facebook and Twitter (see also Vilmer et al. 2018, p. 98).

To return to Chinese investment in Africa, beyond concerns about the BRI's digital component potentially being exploited by authoritarian regimes are those related to BRI projects in general: there are increasing instances of African countries postponing or even cancelling BRI projects over rising debt concerns (Lokanathan, 2020). In a recent working paper commissioned by the Center for Global Development, Landry and Portelance (2021) report that countries such as Angola, Djibouti, and Kenya are heavily indebted to China and that "African countries make up half of the 50 countries most indebted to China as a percentage of GDP" (Landry and Portelance, 2021, p. 3). In addition, Risberg (2019) refers to the "debt trap diplomacy" narrative according to which "China provides infrastructure funding to developing economies under opaque loan terms, only to strategically leverage the recipient country's indebtedness to China for economic, military, or political favor" (p. 43). The narrative suggests that China seizes assets in cases where countries renege on their debt agreements. Research around BRI debt and seizure of assets is, however, mixed. In a special report on the economic impact of BRI on Africa, and specifically on Ethiopia, Kenya, and Nigeria, Adeniran et al. (2021) recount that the benefits appear to be greater than the costs, with potential growth in areas such as trade and commodities. However, they also caution that Chinese BRI investment is resulting in uneven benefits across Africa, with "[t]op commodity producers and exporters" (Adeniran et al., 2021, p. 2) reaping most of the rewards. As for debt traps, Jones and Hameiri (2020) report that the narrative around seizure of assets is a false one, based on accounts that Sri Lanka and Malaysia were victims of China's predatory system in 2017 and 2018, respectively. These researchers point out that "Their debt distress has not arisen predominantly from the granting of predatory Chinese loans, but rather from the misconduct of local elites and Western-dominated financial markets" (Jones and Hameiri, 2020, p. 2).

Whatever public and private views of BRI investments in Africa encompass, the fact remains that the continent's debt burdens hamper the design and development of ICT platforms that are able to accommodate AI technologies. This is in addition to the reality that many countries lack adequate physical infrastructure related to transport, communication, and power to drive AI ecosystems. With regard to the latter, several African nations pay far more for energy than other nations, despite the fact that they might in fact consume very little of it: this reality hits home when one considers that "in Mali, for example, the average person uses less electricity in a year overall than a Londoner uses just to power their tea kettle" (Lakmeeharan et al., 2020, p. 2). With respect to social infrastructure, African nations also face challenges with regard to access to quality water, sanitation, healthcare, and education (cf. Auriacombe and Van der Walt, 2021). Presently, a handful of countries such as Ethiopia, Ghana, Kenya, and South Africa are employing AI-enabled technologies in sectors that include healthcare, agriculture, and finance (Gadzala, 2018), while most African markets have been left out of the AI loop. Infrastructural deficits, coupled to uneven socio-economic development and growth across and within nations, are why we call for a focus on AI that acknowledges that this technology is not the panacea to Africa's challenges – that it needs to be consistently viewed through a socio-technical lens that recognises colonial, post-colonial, and contemporary realities.

MAKING (NON)SENSE OF AI AS A TECHNICAL ARTEFACT

Specific aspects of debates and policy guidelines around AI's design and capabilities do not appear to have been generated directly around public conversations, and yet respect for and inclusion of these conversations is essential with a view to "crafting informed policy and identifying opportunities to educate the public about AI's character, benefits, and risks" (Zhang and Dafoe, 2019, p. 3). Of significance is that public perceptions are to some extent moulded by the mass media, which may exaggerate AI's capacities as well as obfuscate ethical and epistemological issues. Such muddying of the waters, as it were, does little to facilitate substantive deliberation around AI and obstructs informed discussion around how it should be regulated (Cave et al., 2018, p. 4).[1] Even a cursory scrutiny of newspaper headlines across Africa demonstrates that media outlets have a tendency to frame AI in terms of utopian or dystopian rhetoric.

BOTSWANA: "Tech Innovation Enlisted in GBV War" (*Botswana Guardian*, 30 July 2021)
EGYPT: "Egypt to Introduce Artificial Intelligence in Irrigation Water Management" (*Egypt Today*, 10 August 2020)
GHANA: "Government Urged to Act Swiftly to Prevent 'Killer Robots' Development" (*Ghana News Agency*, 27 August 2021)
NAMIBIA: "AI, Biometrics and No Protection from Abuse" (*Namibian*, 24 February 2021).
NIGERIA: "Machine Learning May Erase Jobs, Says Yudala" (*Daily Times*, 28 August 2017)
SOUTH AFRICA: "Artificial Intelligence Trained to Identify Lung Cancer" (*The Citizen*, 22 May 2019)

AI is omnipresent, embedded as it is in smartphones, chatbots, voice assistants, global positioning systems, spam email filtering, and so forth. Yet few scholars in Africa have closely examined how the media may shape societies' perceptions of AI (Brokensha, 2020; Brokensha and Conradie, 2021; Guanah and Ijeoma, 2020; Njuguna, 2021). In their content analysis of three popular Nigerian newspapers with high circulation rates, and in line with other studies' findings (e.g., Ouchchy, Coin and Dubljević, 2020), Guanah and Ijeoma (2020) note that across all three outlets coverage of AI was fairly superficial, given that reports did not interrogate AI's multiple facets and impacts in any in-depth manner. They conclude that newspapers are obliged to critically appraise AI, contending that "Since automation may be the future, newspapers must start to intensify the education of the public about AI" (Guanah and Ijeoma, 2020, p. 57). An insightful study by Njuguna (2021) of online users' comments generated around East African news outlets' reports on sex robots points to most users perceiving "the robots as 'destroyers' of the God-ordained family unit and tools of dehumanizing women, and thus morally contradictory to Christian teaching" (Njuguna, 2021, p. 382). Studies on media framing of AI in South Africa suggest that, as is the case globally (e.g., Duberry and Hamidi, 2021; Fast and Horvitz, 2017), AI is depicted in dramatic or alarmist terms that are not aligned with reality (Brokensha, 2020; Brokensha and Conradie, 2021). Brokensha (2020) and Brokensha and Conradie (2021) have found that press coverage reflects a tendency to employ anthropomorphic tropes that either stress an AI system's human-like form/social attributes or describe its cognitive capabilities. With respect to the former type of anthropomorphism, these researchers note that a human-like appearance or human-like traits are commonly ascribed to AI-enabled social companions or digital assistants in the areas of human-AI interaction, healthcare, and business and finance. With respect to the latter, and particularly in the context of machine learning and neural networks, journalists to some extent portray AI systems as surpassing human intelligence.

When it comes to developing social robots, anthropomorphic design is not uncommon, given that it goes some way to enabling acceptance of and interaction with robots (Fink, 2012, p. 200; cf. Darling, 2015). While this type of anthropomorphism poses a number of significant and potential problems, such as that related to the perpetuation of heteronormativity (Ndonye, 2019) or the establishment of para-social relationships between users and machines (Boch, Lucaj and Corrigan, 2021, p. 8), we are of the view that cognitive anthropomorphism of AI by the media should at this stage concern us more, one of the main reasons being that African policies and outlooks on technology continue to point to techno-deterministic assumptions at the expense of social context and human agency (Ahmed, 2020; Diga, Nwaiwu and Plantinga, 2013; Gagliardone et al., 2015; Williams, 2019). Writing in the context of journalism research in Africa, Kothari and Cruikshank (2021, p. 29) contend that rather than underscoring technochauvinism, the focus needs to shift to equipping humans with the skills they need to grasp AI's consequences and benefits.

Both in and across *The Citizen*, *Daily Maverick*, *Mail & Guardian Online*, and *SowetanLIVE*, Brokensha and Conradie (2021) established that journalists are disposed to framing AI systems as matching or transcending human intelligence. Thus, typical headlines or claims were those such as "AI better at finding skin cancer than doctors: Study" (*Daily Maverick*, 29 May 2018) (see Brokensha and Conradie, 2021, para. 21) and "A computer programme [...] learnt to navigate a virtual maze and take shortcuts, outperforming a flesh-and-blood expert" (*The Citizen*, 9 May 2018) (see Brokensha and Conradie, 2021, para. 20). Of interest is that in recognising the dimension of uncertainty in emerging technologies, journalists may attempt to mitigate how they frame AI through the use of discursive strategies such as scare quotes and paraphrases/quotations of various actors' voices that effectively frame AI's cognitive capacities in dualistic terms (Brokensha, 2020; Brokensha and Conradie, 2021). What is unfortunate about employing competing frames, however, is that they reflect a false balance (Boykoff and Boykoff, 2004, p. 127) that may in turn make it difficult for readers to make a distinction between reality and falsehood (Brokensha and Conradie, 2021). White and Lidskog (2021) stress that to "outsiders" such as the general public, the nature of AI and its associated risks are generally unfathomable, and a rather sobering thought in this regard is that they may also be fairly incomprehensible to "insiders", such as AI developers and researchers. Dualistic framing aside, conflating artificial intelligence and human intelligence is misleading, conveying the message to the general public that AGI has come to fruition. Framing AI in utopian terms reflects what Campolo and Crawford (2020, p. 1) refer to as "enchanted determinism", which sees technology as *the* solution to all human ills. If machines are perceived as surpassing human intelligence, then enchanted determinism may also exhibit the type of dystopian lens that Crawford (2021, p. 214) argues sees technology as consuming human beings. Both utopian and dystopian views are problematic, as both dismiss the fact that it is human beings who lie behind technology (Crawford, 2021, p. 214). Further, given that many African countries are exhibiting digital colonialism, and in light of the fact that the design and application of AI are simply entrenching disparities that are "colonial continuities" (Mohamed et al., 2020, p. 664), all AI-stakeholders-in-the-loop need to resist an ahistorical view of AI (cf. Crawford, 2021, p. 214).

Of course, it is not the mass media alone that may bombard the general public with messages that AI is arcane or inexplicable, thus creating the impression that it cannot be regulated in terms of design or application and that we face inevitable doom. Atkinson (2016) succinctly captures the dilemma that the public experience when thinking about AI when he claims that "fearful voices now drown out the optimistic ones" (Atkinson, 2016, p. 9). In this regard, "fearful voices" include industry experts and scholars. Significantly, and employing the social amplification of risk framework designed among other things to understand and assess risk perception (Kasperson et al., 1988), Neri and Cozman (2020) have found that public perceptions of the risks around AI are largely shaped by experts who frame this technology in terms of existential threats. With respect to voices from industry, and in the context of a discussion about AI's summers and winters, Floridi (2020) reminds us that "Many followed Elon Musk in declaring the development of AI the greatest existential risk run by humanity. As if most of humanity did not live in misery and suffering" (Floridi, 2020, pp. 1–2). With respect to perceptions of AI by scholars, Atkinson (2016, p. 9) observes that computer scientist Roger Schank made the following comments when Stephen Hawking told the BBC in 2014 that "The development of full artificial intelligence could spell the end of the human race" (Cellan-Jones, 2014): "Wow! Really? So, a well-known scientist can say anything he wants about anything without having any actual information about what he is talking about and get worldwide recognition for his views. We live in an amazing time" (Schank, 2014, para. 3). Alarmist messages result in "mass distraction" (Floridi, 2020, p. 2) that takes us away from the fact that AI-enabled technologies are "normal" (Floridi, 2020, p. 2) and that they can assist us in solving or reducing many of the problems we currently face. "Mass distraction" obfuscates the fact that across the continent we need voices of reason that interrogate the realities and myths of AI in African contexts, and there are several of them in areas such as automation, robotics, finance, agriculture, courts of law, climate change adaptation, trade and commerce, and driverless cars (Famubode, 2018; Magubane, 2021; Mhlanga, 2020; Mupangwa et al., 2020; Nwokoye et al., 2022; Parschau and Hauge, 2020; Rapanyane and Sethole, 2020; Rutenberg, Gwagwa and Omino, 2021; Vernon, 2019; Zhuo, Larbi and Addo, 2021). Studies such as these move beyond the mystique of AI, allowing us to contemplate what we are doing with this technology in the first place. Demystifying AI by shifting attention from it as an existential threat to what it can do for developing countries is key and is what Floridi (2020) calls an "exercise" in "philosophy, not futurology" (Floridi, 2020, p. 3).

Assessing the potential of AI in Africa, Alupo, Omeiza and Vernon (2022) observe that recognising that AI has many benefits is a complicated process that hinges to a large extent on trust as well as on AI innovation, which partly requires that physical infrastructure such as electricity and Internet connectivity be in place. Equally important is that innovation is also contingent on communities adopting this technology and usefully deploying it in a manner that is sensitive to local social norms and expectations, as well as socio-cultural factors. In the next chapter, and in the context of the AI ecosystem in Africa, we consider the importance of decolonising the digital citizen without which innovation cannot take place. We once again reiterate that Africa is not homogenous and so we examine what the digital citizen in Africa could encompass without attempting to presumptuously make generalisations across cultures (cf. Alupo et al., 2022).

NOTE

1 We are not suggesting that media framing of AI is the same in and across countries. As Suerdem and Akkilic (2021) quite correctly assert, media framing differs from cultural context to cultural context. What is more, how media outlets portray AI is dependent on a number of variables such as readership, ratings, and editorial agendas. Further, how the public perceive media framings of AI will hinge on their conceptualisations of AI based on the contexts they operate in.

Chapter 5

Digital citizenship in Africa

Contestations and innovations in the context of AI

INTRODUCTION

The notion of "digital citizenship" emerged in the 1990s in the context of education and was initially defined in fairly unsophisticated terms as comprising both access to technology and the ability to use it in a proficient manner. In an increasingly digitalised world often embedded in unequal social, political, and economic landscapes, and with societies generally perceiving transformative technologies in ambivalent terms, the notion of digital citizenship is a contested and fluid one. Given the constitutive entanglements between humans and technology, a detached view of digital citizenship in relation to AI in Africa is not helpful, and so a useful point of departure when thinking about the qualities of citizens within technological systems is to aim for what Dufva and Dufva (2019) refer to as "an embodied understanding of the underlying structures and dynamics of digital technologies" (Dufva and Dufva, 2019, p. 24). Just some of the structural factors and dynamics that undermine digital citizenship in Africa include as we have noted persistent gender inequalities, ongoing neoliberal tensions, the complex dimensions of the digital divide, the existence of digital dictatorships, and marginalisation of indigenous knowledge systems. Acknowledging and addressing these factors and dynamics allow for an approach that resists the notion that digital citizenship is static and unitary. It is one that in our view aligns well with Henry, Vasil and Witt's (2021) feminist approach, which recognises "the complexity and structural nature of oppression, including the ways that multiple structural inequalities converge [...] to influence experiences of citizenship among differently situated subjects" (Henry et al., 2021, p. 7). Although we interrogate how underlying structures and dynamics shape digital citizens in Africa, we take the lead from Vivienne, McCosker and Johns (2016) by attempting to avoid framing digital citizenship solely in a negative sense to address a variety of problems. Instead, we depict it as a notion that allows for advocacy, innovation, and social change. This approach calls for socio-technical development that, in the words of philosopher Lina Rahm, shifts the focus from a preoccupation with "an individual

DOI: 10.1201/9781003276135-5

aspiration to escape as unharmed as possible" to "trying to change things for collective improvement" (Rahm, 2018, p. 40).

RESPONSES TO THE CURTAILMENT OF DIGITAL CITIZEN ENGAGEMENT

Digital citizenship entails

> being able to find, access, use and create information effectively; engage with other users and with content in an active, critical, sensitive and ethical manner; and navigate the online and ICT environment safely and responsibly, while being aware of one's own rights.
>
> (UNESCO, 2016, p. 15)

While useful, this working definition does not recognise that as is the case in other developing regions, digital citizenship may be a precarious concept on the continent, beset as it is by numerous challenges. One of the major obstacles to fostering digital citizenship engagement in several African nations is its ongoing suppression by authoritarian regimes. In previous chapters, we touched on digital dictatorships that leverage AI-enabled technologies to surveil their citizens and political opponents. Digital dictatorships may also exploit these technologies to reduce or entirely hamper citizen engagement and activism on digital platforms. In this regard, Lise Garbe (2020, p. 31) reports on the commonplace practice of Internet shutdowns by some states on the African continent (cf. Gopaldas, 2019). Referring to Internet closures in countries such as Cameroon, Egypt, Sudan, Togo, and Zimbabwe, Garbe (2020) observes that reducing admission to specific websites or obstructing Internet access typically occurs during times of political contestation and electoral tension. At the same time, Garbe (2020) notes that Internet shutdowns are not widespread on the continent, with a third of all countries engaging in such action. Reasons why most African countries refrain from shutdowns are speculative; they may revolve around the extensive economic damage they inevitably cause or around the fact that many governments simply do not have direct access to – and thus control over – Internet service provider operations (Garbe, 2020). An additional and far more interesting reason why shutdowns do not occur everywhere or for extensive periods of time pertains to the actions of local or international human rights organisations and civil society groups, two influential movements being #KeepItOn and Article 19 (Garbe, 2020, p. 33). The former is a global coalition group that was launched in 2016 by Access Now, and in line with the qualities of advocacy, participation, and plurality discussed in earlier chapters, aims to halt Internet shutdowns through approaches such as "grassroots advocacy, direct policy-maker engagement, technical support, corporate accountability, and legal intervention".[1] Article 19 is a group that defends the freedom to speak and to know, and "[w]hen either of these freedoms comes under threat [...] speaks with one voice, through courts of law, through global and regional organisations, and through civil society wherever [they] are present".[2] Since governments across Africa may also negatively impact online civic spaces through disinformation and surveillance, a 2021 report commissioned by the Institute of Development Studies (IDS) details a number of strategies that could empower citizens to challenge and address these abuses with a view to asserting their (digital) rights (Roberts and Ali, 2021, p. 11). Extrapolating data on digital rights from Cameroon, Egypt, Ethiopia, Kenya, Nigeria, South Africa, Sudan, Uganda, Zambia, and Zimbabwe, the IDS report calls for key players that include researchers, journalists, policymakers, lawyers, activists, and the public not only to increase awareness around digital threats to democracy but also to develop the capabilities to effectively address these threats (Roberts and Ali, 2021, p. 33). A practical example of these strategies is highlighted by Karekwaivanane and Msonza (2021) in their IDS report on the digital rights landscape in Zimbabwe. Responses by citizens to the Zimbabwean government's attempts to surveil and control online engagement during times of socio-political upheaval include establishing hashtag campaigns (e.g., #ZimbabweanLivesMatter) to foster cyber-activism and creating awareness programmes to sensitise citizens to the benefits of communicating their protest actions through more secure applications such as Signal and Telegram (Karekwaivanane and Msonza, 2021, p. 53). While the latter strategy is regarded as particularly effective and has also been employed in countries that include Egypt and Sudan, it has not been widely adopted, possibly because few laypeople are technologically aware (Roberts and Ali, 2021, p. 28). The IDS report therefore recommends that all stakeholders work together to increase awareness of digital security tools such as encrypted virtual private networks or VPNs. These networks allow citizens and activists to gain access to social media platforms that have been blocked as well as to maintain communication with one another in the event that the Internet becomes inaccessible (Xinwa, 2020, p. 6). To return to the notion of embodied understandings of digitality, Dufva and Dufva (2019, p. 22) have conceptualised "digi-grasping" as a way for communities to employ digitalisation, not only as a form of self-expression but also as a means to achieve significant change through action. Both cyber-activism in the form of hashtag campaigns and awareness and use of VPNs reflect this concept, allowing communities to employ digitality in attempts to bring about change. Although communities might not necessarily know how the technologies they use work in a technical sense, "there is a feeling, an embodied vision of what direction to take" (Dufva and Dufva, 2019, p. 23). Within Henry et al.'s (2021) feminist framework, collective performativity of civic duty in online spaces that reflect engagement, action, and responsibility is a critical aspect of digital citizenship.

RESPONSES TO UNEQUAL DIGITAL ACCESS AND (RESPONSIBLE) DIGITAL LITERACY

In order to communicate dissent and effect changes for good, communities should be afforded the same rights within AI and other technological ecosystems as they should be offline, benefitting from qualities of digital citizenship that include agency, participation, empowerment, and democratisation, for example (Henry et al., 2021, p. 3). Yet, in and across regions in Africa, unequal digital access and literacy pose significant threats to these qualities, with a recent Afrobarometer Policy Paper constituting a stark reminder of this fact: Krönke (2020, p. 2) reports that in a study of 34 countries in Africa, only 20% of adults enjoy access to a computer or smartphone, with 43% utilising unsophisticated mobile phones. Rudimentary digital literacy is the norm, although significant differences occur from region to region: "In Mauritius, Gabon, Tunisia, Sudan, South Africa, and Morocco, more than half of respondents frequently use cell phones and the Internet. By contrast, the same is true for no more than one in 10 citizens in Mali, Niger, and Madagascar" (Krönke, 2020, p. 8). Two major factors that continue to generate uneven digital access and literacy are a lack of network coverage for mobile phones and limited-to-no access to electricity grids (Krönke, 2020, p. 8; cf. Gadzala, 2018). Since digital access and basic digital literacy challenges are prevalent on the continent, it is not unexpected that access to AI systems and the skills to develop and deploy them are uncommon. In 2018, Gadzala reported that most African countries lack the skills and infrastructure required to establish thriving AI ecosystems and that any projects launched are often pilot studies that tend to be rather arbitrarily implemented (Gadzala, 2018, p. 2). This state of affairs has not significantly improved since then, with Wairegi et al. (2021, p. 14) observing the sluggish pace at which AI is being advanced and deployed when compared to what is presently transpiring in developed regions. Digital connectivity underlines the global sustainable development goals (SDGs) set up by the United Nations General Assembly in 2015 but is currently not resulting in inclusive growth since it remains in the hands of a privileged few while simultaneously ignoring the socio-economic inequalities and inequities considered in earlier chapters (cf. Gillwald, 2020). We argue that in order to reap the benefits of AI, and as a response to digital access and digital literacy challenges, what is called for is a "radical digital citizenship" (Emejulu and McGregor, 2019, p. 131), rather than a focus on access and skills acquisition alone, which is similar to Dufva and Dufva's (2019) notion of digi-grasping. Although Emejulu and McGregor's (2019) framework does not focus on AI *per se*, it is nevertheless useful since it takes into account (1) how use of technologies may have negative political, socio-economic, and environmental impacts on communities as well as (2) how communities may collectively develop and deploy technologies for emancipation and public good. The type of cyber-activism described in the previous section underscores the need not only to be aware of how governments may be exploiting technologies to undermine citizens' political rights and destabilise opponents but also how to harness alternative and secure digital platforms to counter digital authoritarianism. Both are therefore good examples of radical digital citizenship in practice. Emejulu and McGregor (2019) contend that radical digital citizenship should involve shifting from a focus on digital literacy and skills alone to understanding that in a datafied world, citizenship entails "a process of becoming" (Emejulu and McGregor, 2019, p. 140).

Although we recognise the importance of this shift, we nevertheless aver that careful attention needs to be paid to digital literacy in order to establish and maintain citizens' digital rights (Roberts and Ali, 2021, p. 17). We are not suggesting that the acquisition of digital literacy be confined to a type of instrumental literacy however (Pangrazio and Sefton-Green, 2021). In line with Ng's (2012) digital literacy model, what we are calling for is the type of literacy that transcends the technical component to also include its cognitive and socio-emotional dimensions. The cognitive component requires the ability to activate critical thinking and reasoning skills when using or appraising digital data (Ng, 2012, p. 1068), and so within the AI ecosystem, this component would reflect the capabilities necessary for critically interrogating AI's development and deployment. More importantly in our estimation is the inclusion in this component of ethical digital literacy (Ng, 2012, p. 1068), which, in the context of AI, would reflect citizens' knowledge of this technology's ethical and moral facets, as well as their ability to use it in socially responsible ways. After all,

> ...it is not the AI artefact or application that is ethical, trustworthy or responsible. Rather, it is the people and organizations that create, develop or use these systems that should take responsibility and act in consideration of human values and ethical principles [...].
>
> (Dignum, 2021, p. 3)

Writing from the perspective of Sub-Saharan Africa, Gwagwa and his colleagues remind us that responsible AI enables all stakeholders to prosper (Gwagwa et al., 2021, p. 5). When conceptualising responsibility in AI in the context of digital citizenship, we extrapolate recommendations from Henry et al. (2021) who, among other things, advocate the type of transformation that empowers individual citizens to use technology in ways that are ethical, safe, as well as appropriate, and that avoid what they describe as "the universalized, depoliticized, individual citizen-subject" (Henry et al., 2021, pp. 4–5). Such avoidance is an important reminder that digital citizenship is multifaceted and that it cannot be viewed in a vacuum, divorced from any given region's socio-cultural and socio-economic contexts (Njenga, 2018, p. 4).

Intersecting with the cognitive dimension is the complex socio-emotional or human skills component, which, in terms of Ng's (2012) definition, stresses individuals' ability to comprehend and appropriately apply the "rules" associated with navigating cyberspace. Applied to AI settings, this crucial dimension is both under-developed and ill-defined – particularly in the context of vulnerable as well as at-risk populations – requiring urgent attention in education, the workplace, and the community. Helpfully, Angela Lyons and her colleagues have identified key components of this dimension in the era of AI, which are "(1) self-awareness, (2) basic cognition, (3) higher-order cognition, (4), communication ability, (5) social awareness, (6) self-management, (7) professional skills, and (8) personality traits" (Lyons et al., 2019, p. 7). It goes without saying that these soft skills cannot be underrated in the age of the fourth industrial revolution: "With the rise of emerging technologies, such as artificial intelligence, machine learning, and automation entering our workforce, the future of employment will necessitate soft skills that machines cannot replace" (Karimi and Pina, 2021, p. 23). What is more, soft skills may even be described as "the new hard skills" (Dolev and Itzkovich, 2020, p. 55) necessary to manage the uncertainty and disruption that AI is currently generating in education, research, and the workplace.

In essence, digital citizenship – and of course accompanying digital rights – pivot around the technical, cognitive, and socio-emotional dimensions. In addition, as Pangrazio and Sefton-Green (2021) concisely put it, "digital literacy is perhaps foundational for digital citizenship and digital rights: individuals cannot participate or claim their digital rights if they are not 'literate' in the first place" (p. 21).

RESPONSES TO DIGITAL CITIZENSHIP THROUGH DIGITAL EDUCATION AND RESEARCH

Problematically, while digital education and research constitute key solutions to fostering digital literacy and digital citizenship in Africa, how the latter has been (and continues to be) theorised remains elitist in the sense that it tends to be constructed in hegemonic terms, essentially encompassing Global North citizens who are predominantly white, male, affluent, and well educated (Emejulu and McGregor, 2019, p. 139). As far back as 2014, Emejulu put it aptly when she observed that "the focus and fetishisation of the 'new' in relation to digital technologies has crowded out discussions of politics and power" (Emejulu, 2014, para. 2).

The dismissal of such discussions also triggers the negation of digital education and research that reflect communities' indigenous knowledge systems that are intimately intertwined with their lived experiences and realities. Returning to a socio-technical lens on AI and acknowledging that decolonised education and research can be separated neither from politics and power nor from a plurality of worldviews, we support Mohamed et al.'s (2020) emphasis on ensuring that marginalised knowledge systems are given a legitimate platform within AI ecosystems. Conceptually, indigenous knowledge has been defined in several ways by scholars in a variety of fields (Hart, 2010; Horsthemke, 2008; Johnson and Mbah, 2021), but there is general consensus that it resists exclusively Eurocentric or Western-centric epistemologies and rests on communal understandings of knowledge specific to a given region (Ezeanya-Esiobu, 2019, pp. 7–8).

Focusing on African higher education and research in the context of disruptive technologies such as AI and robotics, curriculum studies scholar Kehdinga (2020) points to the continued illegitimacy of African knowledge systems as a result of a colonial legacy steeped in Western paradigms and traditions. Indeed, in a historical sense, indigenous knowledge systems are deemed in academe and education to be inferior to Western science, frequently described "as 'primitive', 'backward', 'savage', 'rural', 'unscientific', and so on" (Ezeanya-Esiobu, 2019, p. 7). The prevailing attitude, in other words, is that such systems cannot be combined with scientific objectivity – that they simply have no epistemic value, given that they are generated on the basis of communities' lived experiences (Tao and Varshney, 2021). Kehdinga (2020) warns that a sustained focus on Western knowledge as superior to other forms of knowledge "creates a deep rooted and normalized epistemological enclosure, which prevents the emergence of critical African knowledge to circulate in any substantive way" (Kehdinga, 2020, p. 250).

Be it in the field of education or in the domain of research, both indigenous knowledge and Western science "are ultimately based on observations of the environment, both provide a way of knowing based on these observations, and both emerge from the same intellectual process of creating order out of disorder" (Berkes and Berkes, 2009, p. 8). However, Western knowledge is constructed in such a way that it is to some extent divorced from analyses of human constructs and dismissive of historicity and ideology (Paraskeva, 2011). Western science's focus on rationality and calculation "makes knowing a distant act" (Birhane, 2021, p. 2). Yet as we have seen, aspects of AI's design and implementation may be "messy" and unethical: AI companies could exploit low-skilled workers in the gig economy; contrary to perceptions or expectations, AI systems do not represent the elixir for Africa's problems; and machines used in the healthcare and business sectors, for example, may mirror biased patterns as a result of inaccurate or unrepresentative training data (cf. Chan et al., 2021; Mhlambi, 2020; Singh, 2019). The latter problem is an especially good example of Western epistemologies' limitations: in terms of Western orthodoxy, biases may be resolved merely by applying measurements to attain algorithmic fairness. Such rational logic ignores the fact that data collection may not necessarily take place within just systems (Chen, Joshi and Ghassemi, 2020, p. 16) and that the indigenous knowledge systems of communities affected by algorithmic discrimination also need to be taken into account in AI data training (cf. Chan et al., 2021, p. 4).

Kehdinga (2020) proposes that the legitimacy of indigenous knowledge may partly be established through what Paraskeva (2011) refers to as an itinerant curriculum, which is a response to epistemicide, defined as "the killing, silencing, annihilation, or devaluing of a knowledge system" (Patin et al., 2021, p. 1306). A critical itinerant curriculum is one that is geared towards cultivating social justice and equality while resisting curricula based exclusively on Western epistemologies (Paraskeva, 2011, pp. 16–18). What is not in dispute is the need for quality STEM education on the continent to prepare its citizens for the fourth industrial revolution, but in the context of itinerant curriculum theory, knowledge and skills in STEM need to be enhanced by the digital literacy dimensions proposed by Ng (2012) and others (Kehdinga, 2019; Karimi and Pina, 2021) that are consistent with Africa's social, political, and economic realities. In addition, there is an urgent need to employ African knowledge in sub-fields of AI education and research (cf. Kehdinga, 2020, p. 250). At present, there are several cases of AI applications in Africa that encourage the amalgamation of African knowledge and technology. In the area of natural language processing (see Chapter 7), for instance, a number of African scholars have begun arguing for the importance of developing AI/machine learning technologies to support indigenous (and often resource-scarce) languages (e.g., Du Toit and Puttkammer, 2021; Marivate, 2020; Shikali and Mokhosi, 2020). Although controversial for reasons discussed in previous sections of this book, facial recognition software employed in Africa has benefitted from input by scholars such as Ivorian researcher Charlette N'Guessan, who, in collaboration with her colleagues, has eschewed Western-designed systems to develop Bace API, which has been designed to accommodate people with dark skin tones (De Vergès, 2021). The coalescing of scientific knowledge and indigenous knowledge is particularly evident in studies that focus on climate change (drought), agriculture, and water management in Africa. Nyetanyane and Masinde (2020), for example, have proposed integrating machine learning, climate data, and satellite imagery with indigenous knowledge with a view to predicting favourable weather patterns for crop cultivation in South Africa, while Thothela et al., (2021) recently evaluated to what extent integrating indigenous knowledge systems into science and technology (such as smart sensors and mobile phones) may assist small-scale farmers in Sub-Saharan Africa to make informed decisions about their crops. Agricultural and aquatic ecosystems as well as aspects of climate change are complex and unpredictable. The absence of boundedness means that positivistic science cannot easily cope with these systems (Berkes and Berkes, 2009, p. 6). Indeed, in the context of socio-cultural landscapes that are rarely orderly and predictable, Svensson (2021) notes that AI systems "do not acknowledge not-knowing, and as a consequence are incapable of dealing with the unknown" (Svensson, 2021, p. 5). We are not suggesting that Western and indigenous knowledge systems are disparate and that the latter should simply replace Eurocentric or Western-centric epistemologies. However, indigenous knowledge reflects a holism that allows it to accommodate natural systems that are both unpredictable and uncertain (Mazzocchi, 2006, p. 466; cf. Berkes and Berkes, 2009, p. 6).

The aforementioned examples of the integration of Western and traditional knowledge systems point to the necessity for digital citizens to be aware of the need *for appropriate technology*, defined by Ezeanya-Esiobu (2019) as the type of technology that goes beyond applying scientific knowledge to also understanding the unique environment in which that technology is used, thus "bringing about transformation and noticeable progress" (Ezeanya-Esiobu, 2019, p. 5). Of equal importance, and drawing on Basu and Weil's (1998) appropriate technology theory, Ezeanya-Esiobu (2019) stresses the development of *localised innovation*, according to which it is not sufficient to merely transfer a particular technology to a foreign environment: "appropriate technology needs to be situated in the preexisting technological knowledge or environmental reality of the innovator. This is where indigenous knowledge comes to the fore" (Ezeanya-Esiobu, 2019, p. 6).

At this stage, it is important to note that stressing an indigenous knowledge system's holistic dimension does not imply that such a system is unscientific and that it should therefore be compared in an unfavourable sense to Western scientific knowledge (cf. Bruchac, 2014; Knopf, 2015; Mazzocchi, 2006; Ogungbure, 2013). It is worth quoting Knopf (2015), who argues that "grounded in holism and relational views", indigenous knowledge "defies categorization as it operates, to speak with Western categories, in an interdisciplinary and transcultural mode" (Knopf, 2015, p. 182). The inference here is that it is counter-productive to regard Western and indigenous epistemologies as dichotomous (Mazzocchi, 2006, p. 464). With respect to developing engaged digital citizens within AI ecosystems, one strategy for "bring[ing] the knowledge systems closer together" (Knopf, 2015, p. 183), albeit a controversial one, is through citizen science, which we briefly discuss at the end of the next section.

PROMOTING DIGITAL CITIZENSHIP IN AFRICA IN THE CONTEXT OF AI: MOVING BEYOND A SYMBOLIC TERM

By adopting a post-colonial lens that recognises many African nations' colonial histories and anti-colonial struggles, we are calling for the development of *critical* digital citizens through decolonial approaches in communities and societies, in research, and in formal/informal education at all levels that encourage critical inquiry into how AI intersects with realities that cannot be placed in neat boxes (Svensson, 2021, p. 1). To avoid dealing with the notion of the digital citizen as abstract (cf. Oyedemi, 2020), we have provided practical suggestions as to how AI may be approached and critiqued in order to bring about transformation as well as sustainability. The proposals recommended apply to all citizens, whether they are consumers or innovators of AI-enabled technologies.

We cannot gloss over the fact that defining "decolonisation" involves much indetermination and dispute, given that scholars' perceptions of what it entails are diverse (e.g., Behari-Leak, 2019; Clapham, 2020). Mbembe (2015), for example, does not entirely negate employing knowledge systems from the Global North, while Behari-Leak (2019) describes decolonisation as a movement that recentres "that which is sacred and indigenous and home-grown" (Behari-Leak, 2019, p. 60). Following Sibanda's (2021) broad definition, it is "largely and essentially a knowledge and power project, meant to redress power differentials in the determination of what is knowledge, its creation, dissemination and evaluation" (Sibanda, 2021, p. 183). In the areas of both AI education and research, we call for a decolonial lens, which perceives (African) knowledges as going beyond abyssal thinking to include "lived knowledges" (De Sousa Santos, 2018, p. 2) that reflect myriad ways of knowing (De Jaegher, 2019, p. 847; cf. De Sousa Santos, 2018, p. 2). Revisiting the relational ethics of care framework considered in Chapter 3, "knowing" on the part of digital citizens becomes what Birhane (2021, p. 4) in the context of Afro-feminist thought describes as "an active and engaged practice" within societies and communities inhabiting specific regions. On a practical level, ways of knowing would include recognising that many of AI's ethical and foundational principles are dominated by Western reason and thus not aligned with Africa's problems, challenging algorithmic coloniality (Mohamed et al., 2020) or understanding the need for localised AI innovation in areas such as agriculture, financial services, and the healthcare sector.

Such a perspective implies a focus on pluriversality, which, as a concept derived from decolonial theory, is not without its problems: we may refer to "the interactions of various epistemic formations under pluriversality" (Gallien, 2020, pp. 28–29) that obviously reflect diverse concepts, methodologies, and languages that invariably bring about social change that is beneficial to some and yet harmful to others. One solution to the problem lies in being careful not to separate decolonisation of AI education or research from context: "context matters, [and] so does the relationality of the diverse aspects of decolonial projects" (Dube, 2021, p. 74). Birhane's (2021) conceptualisation of knowing referred to earlier reflects this sentiment: knowing is socially embedded and constructed in such a way that it embraces the unique contexts in which AI is designed and used. She refers specifically to "enactive cognitive science", a theory which subordinates reasoning to "engaged, active, involved, and implicated knowing"

(Birhane, 2021, pp. 4–5) and is thus aligned with an embodied understanding of AI (cf. Dufva and Dufva, 2019).

Our suggestions for fostering digital citizenship are to some degree consistent with those proposed by Zembylas (2021), who explores the decolonisation of AI in tertiary education settings but with a particular emphasis on its ethical component. Pointing to research conducted by Adam (2019) on the perpetuation of neocolonialism via Western-dominated massive open online courses (MOOCS), Zembylas (2021) offers the caveat that a decolonial approach to AI ethics that includes marginalised stakeholders is not an automatic solution to preventing Western hegemony over (AI) technologies. Indeed – and paradoxically – the inclusion of marginalised voices could result in what Adam (2019, p. 365) refers to as "adverse incorporation": Elwood (2021) cautions that this type of incorporation "implicitly leaves intact the prevailing formulations of digital insiders/outsiders, haves/have nots, or center and margin that have confined us to reproducing existing (profoundly unequal) worlds" (p. 212). In the contexts of higher education and the development of critical digital citizenship, a critical strategy for countering adverse incorporation entails educators and researchers helping digital citizens become skilled in considering how AI-enabled technologies' ontologies and epistemologies could be reimagined to challenge those that are Western-centric in design (Zembylas, 2021, p. 2). On an operational level, the recommendations made here (and indeed throughout this book) echo Zembylas' (2021) practical socio-technical strategies, which encompass "historicizing AI and digital technologies as affective, material and political assemblages of coloniality and racism" (Zembylas, 2021, p. 9).

Our suggestion here is that researchers should be encouraged to examine how integrating populations' knowledge systems into AI – and thus encouraging engaged citizenship – could be enhanced through citizen science, which entails the public becoming directly involved in scientific research endeavours (Vohland et al., 2021, p. 1). The value of citizen science is that it allows people from all walks of life to challenge the notion that "academic knowledge is objective" (Heinisch et al., 2021, p. 112).[3] It is also a way to directly share information about AI with local communities and provide them with informal AI education (Phillips et al., 2019). A practical example of deploying citizen science is documented in a recent study by Torney and his research team, who counted aerial imagery of wildebeest populations in Serengeti National Park in Tanzania. Comparing citizen science and deep learning techniques, they report that the use of both techniques resulted in high levels of accuracy, with citizen science volunteers playing a critical role in the generation of training data (Torney et al., 2019).

Although certainly acknowledged here, it is not within the scope of this book to fully consider scholars' scepticism towards citizen science, to detail its myriad positive and negative dimensions or to interrogate the fact that some of its critics hold the view that it is tied to neoliberalism; these issues and others are identified and critically addressed at great length and in depth in *The Science of Citizen Science* (Springer), edited by Vohland et al. (2021) as well as by several other scholars (e.g., Phillips et al., 2019; Walker, Smigaj and Tani, 2021). Suffice it to say, the way in which citizen science is conceptualised and implemented by scientists may result in the type of adverse incorporation referred to earlier: given that citizen science is (1) based on volunteers coming forward to participate in research projects and (2) frequently overseen by foreign companies and/or the scientific community, it carries the risk of creating significant power imbalances, exploitation, and lack of diversity (Ceccaroni et al., 2021; Guerrini et al., 2021). Historically speaking – and with few exceptions – citizen science has not resulted in significant democratisation of science (cf. Mueller, Tippins and Bryan, 2012). A recent study by Weingart and Meyer (2021) examining 56 South African projects in the life science fields that utilised citizen science shows a concerning disparity between the positive rhetoric around this notion and what occurs in reality: the benefit to citizens who participate in projects is generally hampered by a lack of training in digital literacy and STEM disciplines, local problems are not necessarily addressed through the projects, and citizen scientists often make peripheral contributions that do not allow them to make decisions at the level of science policy. Further, results tend to be shared with scientific communities and not with the general population itself (Weingart and Meyer, 2021, p. 616). To democratise collaborative efforts between scientists and citizen scientists in the field of AI research, it is necessary to put strategies in place to avoid citizens becoming "mere laboratory grunts" (Barton, 2012, p. 1). Ceccaroni et al. (2021, p. 233) stress the need to ask whether citizens' indigenous knowledge systems are indeed respected during the course of a given research project and to interrogate who eventually benefits from the results (Ceccaroni et al., 2021, p. 233). Although Barton (2012) in a response paper to Mueller et al.'s (2012) article on the future of citizen science focuses on youths' engagement with science, she significantly reframes citizen science as "citizens' science", which is a novel way to reimagine citizens' participation in AI projects. In this reconceptualisation, Barton (2012) stresses the importance of three elements, perceiving citizen scientists "as community science experts, individuals with a collective expertise characterized by a deep connection to place, the capacity to use this connection to engage community members, and the knowledge of scientific processes to take action on local issues" (Barton, 2012, p. 3).

We turn now to a discussion of the impact AI is currently having on labour markets in Africa, taking into account what kinds of principles should be underscored when attempting to create workspace conditions that are fair and just.

NOTES

1 https://www.accessnow.org/keepiton-faq/
2 https://www.article19.org/about-us/
3 We are distinguishing here between the natural sciences and social sciences, where the former, as opposed to the latter, reflects higher degrees of objectivity and certainty (see Nakkeeran, 2010).

Chapter 6

AI and the labour sector in Africa

Disruptive or transformative?

INTRODUCTION

Sobering statistics reveal that approximately 34.5% of the South African workforce is currently unemployed (Stats, 2022, p. 8) – the highest unemployment rate on the continent – with at least 32 other African countries poised above 6% (Saleh, 2022). Painting an even bleaker picture is the fact that Africa's youth unemployment rate stands at 13%, while the female unemployment rate (at 9%) is higher than that for males (at 7.4%) (Saleh, 2022). A staggering 39% of women in Djibouti are without employment and in South Africa, joblessness among women hovers at 36% (Saleh, 2022). While significant strides in the context of fourth industrial revolution education have been made on the continent in terms of growth in educational programmes and increased resources in the higher education sector in the aftermath of Western colonialism, disparities persist. Indeed, it remains a fact that inequality and exclusion in education have not been resolved, while high-quality teaching and learning are not the norm.

The future poses further challenges for career opportunities, given that the fear exists that AI, robotics, and automation have the potential to disrupt various occupations (Acemoglu and Restrepo, 2017; Arntz, Gregory and Zierahn, 2016; Bakhshi et al., 2017; Bertani et al., 2020; Bruun and Duka, 2018; Brynjolfsson and Mitchell, 2017; Cabrales et al., 2020; Frey and Osborne, 2017; Ford, 2013; McClure, 2018; Naudé, 2021; Nedelkoska and Quintini, 2018; Parschau and Hauge, 2020; Walsh, 2018; Zemtsov, 2020). Acemoglu and Restrepo (2017), for example, argue that automation can lead to increased unemployment and a reduction in salaries and that the consequences are particularly dire in manufacturing, routine blue-collar, and related occupations, as well as for workers with a lower level of training. Given the significant threat that automation holds for unskilled and semi-skilled labour and the lesser risk it poses for skilled labour – coupled with the increased remuneration and job opportunities associated with higher qualifications – the ideal is for a larger section of the population to

DOI: 10.1201/9781003276135-6

be able to better qualify themselves. One important question, however, is this: what skills will be necessary in the future labour market in which AI is pervasive? An even more important question revolves around what kinds of tenets should be put in place to foster working conditions that are fair, equitable, and just (cf. Cole et al., 2022).

This chapter discusses the impact of AI on the global labour market, with specific reference to Africa. AI's influence on unskilled, semi-skilled, and skilled labour is interrogated as are optimistic and pessimistic views of AI's disruptive and transformative potential. Against this background, the chapter then considers the kinds of tenets that should be put in place when attempting to create workspace conditions on the continent that are fair before concluding with a consideration of the type of education that could mitigate AI's disruptive consequences.

ARTIFICIAL INTELLIGENCE AND THE DISRUPTION OF THE LABOUR MARKET

Semi-skilled labour

Automation's principal impact thus far has been on semi-skilled labour, while unskilled and professional occupations have been fairly resistant to automation (Autor and Dorn, 2013, p. 1559; Ford, 2013, p. 2; Frey and Osborne, 2017, p. 259; Kaplan and Haenlein, 2020, p. 46; Nedelkoska and Quintini, 2018, p. 21). This phenomenon is referred to as job polarisation, which refers to the erosion of middle-income, middle-skilled jobs (Bissessur, Arabikhan and Bednar, 2020, p. 409; Frey and Osborne, 2017, p. 258; Naudé, 2021, p. 7; Nedelkoska and Quintini, 2018, p. 25). AI continues the trend established by earlier forms of automation by targeting semi-skilled positions in particular. Occupations that include many routine tasks, such as bookkeeping, clerical duties, manufacturing, and monitoring, among others, are particularly susceptible to automation (Autor and Dorn, 2013, p. 1559; Frey and Osborne, 2017, p. 255; Zemtsov, 2020, p. 727). In medical practices, tasks such as insurance claims, pre-authorisations, appointment reminders, invoicing, data reporting, and analysis are already automated (Lin, Mahoney and Sinsky, 2019, p. 1628). AI-enabled technologies can also aid retail banks with respect to automating key processes that include home loan applications and customer services (Lee and Shin, 2020, p. 158).

As discussed in this book, the threat of automation in Africa is also gendered, given that women are more likely than men to work in clerical and administrative positions, which are at the highest risk of being automated (UNESCO, 2020b, p. 4). Losing these jobs to AI could entail increased unemployment for women, which could lead to increasing dependence on male family members for financial security.

Unskilled labour

Unskilled labour was previously less susceptible to automation because these tasks were simply more difficult to automate (Nedelkoska and Quintini, 2018, p. 25). Ford (2013) writes, however, that the impact of AI on the labour market is greater than previous automation processes. In fact,

> Robots are rapidly advancing while becoming less expensive, safer, and more flexible, and it is reasonable to expect they will have a potentially dramatic impact on low-wage service sector employment at some point in the not too distant future.
>
> (Ford, 2013, p. 3; see also Autor and Dorn, 2013, p. 1559)

Technology is becoming more widely available and unskilled labour, such as cleaning and gardening can, for example, already be partially replaced by widely available robotic vacuum cleaners, mops, and lawnmowers. It remains a challenge to replace all of a worker's functions (Bruun and Duka, 2018, p. 5), but robots such as Baxter have shown that technology can already perform basic manual labour after only minimal training (Frey and Osborne, 2017, p. 260), although it should be noted that Baxter's manufacturer, Rethink Robotics, ceased operating in 2018 owing mainly to a shortage of funding. Nedelkoska and Quintini (2018, p. 51) have found that unskilled occupations such as food preparation at fast food chains, cleaning, and labour in mining, construction, transportation, manufacturing, garbage removal, fishing, agriculture, and forestry are at the greatest risk of being replaced by automation (see also Frey and Osborne, 2017, p. 261).

Until now, retail was a refuge for unskilled workers who could work as shop assistants and cashiers. However, online sales platforms are making shopping assistants and cashiers redundant (Ford, 2013, p. 3), while customer queries are now regularly answered by bots. Even when it comes to physical stores, Weber and Schütte (2019) contend that functions such as marketing, purchasing, warehousing, record keeping, customer service, and the like can be enhanced by AI.

As with semi-skilled labour, women in Africa are often employed in unskilled positions such as cashiers, shop assistants, fast food workers, and cleaners. In addition, unskilled labourers are the lowest income group and the group with the lowest qualifications among the employed, meaning that unskilled labourers have very few alternative sources of income. The lowest income groups are therefore the most vulnerable, whereas the highest income groups are the least exposed, as discussed in the next sub-section.

Skilled labour

At the other end of the training spectrum, AI can currently perform a variety of tasks that were not previously considered possible to automate. We will mention only a few examples here.

In psychology, conversational agents are used instead of or in addition to a physical psychologist (Bendig et al., 2019). Conversational agents create the opportunity to reach people who would not otherwise be reached, or can be used in addition to visits to a psychologist to enhance the effectiveness of the treatment. Currently, however, conversational agents are used in a supplementary role and do not risk replacing psychologists.

In the medical field, AI is already utilised in diagnostic workups (Lin, Mahoney and Sinsky, 2019, p. 1628). IBM Watson, for instance, claimed in 2016 that it can combine big data and AI to outperform oncologists in complex cognitive tasks, such as cancer diagnosis and treatment recommendations (Frey and Osborne, 2017, p. 259; Kaplan and Haenlein, 2020, p. 43; Nedelkoska and Quintini, 2018, p. 21). Even surgeons' work is partially performed by robots that help carry out surgical procedures (Frey and Osborne, 2017, p. 260). As with other skilled jobs, however, AI is rather used to enhance than to replace job roles as Lin et al. (2019) argue: "We believe that the optimal role of AI is to free up physicians' cognitive and emotional space for their patients" (Lin et al., 2019, p. 1629).

In the legal profession, lawyers and clerks are partly being replaced by software that determines which documents are relevant to a court case (Ford, 2013, p. 3; Frey and Osborne, 2017, p. 259). As mentioned earlier, AI is already employed in law firms in Africa, notably by those located in Kenya, Nigeria, South Africa, Tanzania, and Uganda (Adeyoju, 2018). Markovic (2019, p. 349), however, is of the view that AI is likely to benefit the legal profession by freeing attorneys from routine tasks such as document review that will in turn enable them to focus on their core duties, namely advising clients.

Some forms of journalism have already been replaced by AI (Ford, 2013, p. 3; Nedelkoska and Quintini, 2018, p. 34). Quantitative forms of journalism that reflect news about financial markets, sporting events, the weather, and the like are easier to automate than in-depth reporting, but Liu and his colleagues discuss how Reuters' automated news writer, Tracer, has the capability to handle more difficult tasks (Liu et al., 2017).

As is the case with various skilled professions, automation threatens not so much the profession as it creates opportunities to work more cost-effectively (Lindén, 2017, p. 72).

Translation used to be a specialised profession that required thorough knowledge of source and target languages, but machine learning has put translations through, for example, Google Translate, within everyone's reach (Frey and Osborne, 2017, p. 259). While Google Translate's accuracy varies between language pairs, it is most accurate between European languages, although recent efforts have been made to improve the accuracy of Google Translate for low-resource languages (Caswell and Liang, 2020). Vieira (2020, p. 7) does however point out that translators will adapt to AI by, for instance, moving away from technical translations to more creative ones that AI cannot do.

Despite the fact that some skilled occupations are at present being disrupted or transformed by AI, Nedelkoska and Quintini (2018, p. 51) aver that teachers, managers, and other professionals are at the lowest risk of their occupations being automated. In fact, they argue that the threat of automation decreases as the required skill levels for a profession increase: "Educational attainment shows a very clear pattern in relation to automatability: a higher educational attainment translates into a lower risk of automation" (Nedelkoska and Quintini, 2018, p. 53; see also Zemtsov, 2020, p. 732). In addition, those in highly skilled professions are in a better position to adapt to changing circumstances: that is, "workers in high-skilled occupations may have greater ability to learn new information, tend to possess skills which cannot be easily automated and have greater access to lifelong learning" (Nedelkoska and Quintini, 2018, p. 35).

Furthermore, professions that have a greater degree of creativity, communication, empathy, entrepreneurship, and skills that help people work with advanced technology, such as software engineering, are generally more difficult to automate (Naudé, 2021, p. 7; Nedelkoska and Quintini, 2018, p. 25). It may also be the case that automation enriches skilled professions rather than replaces them (Nedelkoska and Quintini, 2018, p. 25; Zemtsov 2020, p. 740). In other words, while AI and automation threaten unskilled and skilled labourers, it constitutes less of a threat to skilled professionals.

A pessimistic view of the impact of automation on the labour market

Ford (2013) does not expect the majority of people to be able to adapt to AI's disruption to the labour markets:

> If we assume [...] a normal distribution of capability among workers, then 50% of the workforce is by definition average or below average. For many of these people, a transition to creative/non-routine occupations may be especially challenging [...].
>
> (Ford, 2013, p. 2; see also Zemtsov, 2020, p. 729)

Bruun and Duka (2018) are by implication even more pessimistic about people's ability to adapt to these rapid technological advances when they propose that all citizens should receive a basic allowance with income recovered from taxes paid by automated firms. This means that the authors expect large-scale unemployment that can only be addressed by a universal grant.

One of the major risks of automation is the influence it can have on societies already impacted by the disparities considered throughout this book.

Technological advances are driven by those who have had the opportunity to qualify themselves in the most suitable disciplines, and job security is therefore best guaranteed for those with higher qualifications and higher income levels. As noted earlier, this is the domain of mostly white affluent males who possess technical skills (West et al., 2019, p. 6). In Africa, on the other hand, many women, the economically vulnerable, and marginalised groups, do not hold secondary or tertiary qualifications and are thus threatened by job automation. Young women are also at risk of losing their jobs to automation, given their socio-economic and lived experiences. Among other things,

> In Africa, young women are over 1.5 times less likely than young men to be formally employed or undergoing education or training. Unequal access to educational opportunities, early marriage rates among young women, and responsibilities for unpaid care and housework are some of the factors contributing to this disparity.
>
> (Santos and Rubiano-Matulevich, 2019, para. 1)

These dynamics unfortunately create the opportunity to widen the gap between rich and poor and further divide society even more than it currently is Bissessur et al. (2020, p. 409). Zemtsov (2020) maintains that

> We are talking about the formation of a group of people, using pension and social benefits as a source of income, as well as leading a subsistence economy. This will be a sector of shadow employment, semi-legal part-time jobs with [...] low living standards, and high mortality rates.
>
> (Zemtsov, 2020, p. 739)

Naudé (2021, p. 8) also highlights the fact that technological advances can widen the gap between developed and emerging countries and that the labour market of the latter will particularly be disrupted by automation. As argued elsewhere in the current book, this view suggests that AI has the potential to further entrench patterns of privilege and deprivation along racial lines and worsen the unequal resource distribution between Africa and Europe.

Although low-income countries in Africa are susceptible to automation (see Millington, 2017), some researchers are more optimistic. The following sub-section considers a more positive vision with respect to the impact of automation in the future.

An optimistic view of the impact of automation on the labour market

It remains an open question whether AI will not possibly create new careers (Bertani, Raberto and Teglio, 2020, p. 331; Bruun and Duka, 2018, p. 5;

Naudé, 2021, p. 10; Nedelkoska and Quintini, 2018, p. 38, 2021, p. 35; Parschau and Hauge, 2020, p. 128; Sako, 2020, p. 26). Frey and Osborne (2017, pp. 255–258) observe that technological advances over the past five centuries have often been accompanied by fears that this will lead to increased unemployment, whereas the opposite has been the case (Bakhshi et al., 2017, p. 22; Parschau and Hauge, 2020, p. 121; Zemtsov, 2020, p. 726). Nedelkoska and Quintini (2018, p. 38) also state that although the development of new AI concepts and techniques will usually be the domain of those trained at the doctoral level, a much broader landscape of skills is needed to integrate these concepts into technologies and systems.

In addition, AI will have to be more cost effective than humans, which is not the case at present. Indeed, the cost of AI is a major obstacle currently affecting the uptake of this technology (Kaplan and Haenlein, 2020, p. 43; Naudé, 2021, p. 11), a fact that is also a reality in South Africa (Parschau and Hauge, 2020, p. 125). Frey and Osborne (2017, p. 261) do, however, point out that costs associated with technology are declining annually, which will place increasing automation within the reach of more organisations.

There is also the issue of market regulation. Kaplan and Haenlein (2020, p. 46) write that governments may decide to ban automation for the sake of the labour market, but regulation is a short-term solution because it will negatively affect the competitiveness of economies, particularly as this plays out between developing and developed countries.

A number of scholars point out that most professions demonstrate a great deal of internal variation and therefore most professions will be disrupted by automation, although they will not be made entirely redundant either (Bakhshi et al., 2017; Bissessur et al., 2020; Kaplan and Haenlein, 2020; Naudé, 2021; Nedelkoska and Quintini, 2018; Sako, 2020). In such a case, technology is incorporated into existing tasks to unlock human capital for other tasks, which means that technology contributes to a worker's performance. Zemtsov (2020, p. 727) also writes that most professions will continue to exist despite automation, with stenography being a recent exception as this profession was officially removed as a profession in Russia in 2018.

In our view, AI will in most cases be utilised as a tool to complete tasks within occupations more cost-effectively while occupations as a whole will not be automated. According to this scenario, workers will use AI to execute their jobs more effectively and efficiently, and will also need to adapt to changing circumstances.

AI ETHICS IN THE WORKPLACE

Cole et al. (2022, p. 2) quite rightly point out that in the literature on AI ethics in the workplace, a significant lacuna involves how AI ethics may be operationalised in this space. Partially defining fairness as a virtue that focuses on "increasing individual and organizational capabilities to guide us toward a less unfair society" (Cole et al., 2022, p. 5), these scholars recommend a number of concrete principles that should receive attention and that have the potential to address dynamics of power and control in workspaces. These include fostering explainability (discussed in Chapter 3), improving individuals' understanding of the data risks that AI-enabled technologies pose, reducing the risks of AI destroying jobs, guaranteeing decent labour standards, and including workers' voices in AI debates (Cole et al., 2022, p. 11). To date, the literature on AI ethics in the context of African labour markets is sparse, yielding little discussion around the risks AI poses to the workforce and how these risks could be mitigated (Gaffley, Adams and Shyllon, 2022; Poisat, Calitz and Cullen, 2021). While the relational ethics of care considered in Chapter 3 is useful, it does not speak *directly* to issues of justice, equity, and fairness in the workplace as Cole et al.'s (2002) principles do. In Chapter 2, we interrogated the so-called gig economy, which is increasingly steered by AI and which leaves many workers vulnerable to poor work conditions that include minimum wages and no benefits such as sick leave or access to medical aid. Exploitation of workers in the gig economy is prevalent in Africa, particularly as a result of low unionisation (see Rani and Furrer, 2021, for example). Illustrative tenets such as those proposed by Cole et al. (2022) reflect a shift away from an abstract ethical framework towards one whose operationalisability could help address the concerns of clickworkers on the continent.

THE IMPACT OF AUTOMATION ON EDUCATION AND TRAINING

So far, it is clear that job security is better guaranteed in skilled occupations and less so in semi-skilled and unskilled occupations, although the whole spectrum of the labour market will be unsettled by technology. The rule is that the more routine tasks a profession involves, the more that profession is threatened, although it must be borne in mind that what is regarded today as non-routine work may be within the reach of AI in the future (Bertani, Raberto and Teglio, 2020, p. 330; Ford, 2013, p. 2; Frey and Osborne, 2017, p. 255; Sako, 2020, p. 26). In the early 2000s, clerks, administrative support workers, sales workers, production workers, and repair workers were seen as occupations that could be automated, while the list now includes drivers, translators, tax analysts, medical diagnosticians, legal assistants, security guards, law enforcement officers, human resource personnel, and even software programmers (Nedelkoska and Quintini, 2018, p. 34). Ford (2013) avers that workers will have no choice but to make a shift towards occupations that are non-routine – that will, in other words, not be threatened by automation (see also Kaplan and Haenlein, 2020, p. 47; Zemtsov, 2020, p. 729).

Education systems on the African continent must make provision for this disruption because the great need for administrative and clerical knowledge highlighted, for example, by the South African Department of Higher Education and Training (2019, p. 103), may in future be filled by AI. Bruun and Duka (2018) argue that schooling must take into account this rapidly changing world:

> In the age of AI, computers will be able to [gather information] more quickly and accurately than humans. It is therefore essential that our schools begin to better equip students with the necessary tools to handle work in the twenty-first century.
>
> (Bruun and Duka, 2018, p. 10)

Nedelkoska and Quintini (2018, p. 28) also write that developed economies are moving away from learning routine cognitive and manual tasks to learning non-routine cognitive and interactive tasks.

Computer skills are the key to adaptability (Zemtsov, 2020, p. 729). Nedelkoska and Quintini (2018, pp. 85–86) have found that only 23% of workers with a lower secondary qualification or less use computers in their work, compared to 98% for workers with a degree and 100% for workers with a Master's degree or higher qualification. This means that the more individuals require a computer for their work, the smaller the chance that what they do can be replaced by automation (Nedelkoska and Quintini, 2018, p. 92): expressed a little differently, "computers do substitute labour, but not the labour of those who use them directly" (Nedelkoska and Quintini, 2018, p. 93).

Regardless of the course for which a student registers, computer skills must in other words be part of the course content. More specifically, Kaplan and Haenlein (2020, p. 47) write that every person should have at least a basic understanding of programming and learn a programming language like Python. If learning a programming language is not feasible, every person will need to have at least a basic knowledge of how machine learning works and what can be achieved with it. Lee and Shin (2020) call for managers in particular to familiarise themselves with machine learning:

> [...] managers at all levels need to familiarize themselves with machine-learning techniques to ensure that their portfolio of machine-learning projects – either in operation or under development – creates maximum value for their enterprises.
>
> (Lee and Shin, 2020, p. 158)

As noted in an earlier chapter, Karimi and Pina (2021, p. 23) argue that with the introduction of new technologies like automation, machine learning, and AI into our workforce, the demand for soft talents that computers cannot replace will increase. In addition to computer skills, reading, writing, and learning skills will play an increasingly important role in the future labour market, as highlighted by Bakhshi et al. (2017) and the Department of Higher Education and Training (2019, p. 103). Rios et al. (2020, p. 83) also found that the skills most in demand are oral communication, written communication, collaboration, and problem solving (Rios et al., 2020, p. 83). Kosse and Tincani (2020) have demonstrated how important social skills are in the workplace: in a global study, they concluded that social skills are correlated with income levels and that social skills are negatively correlated with unemployment. Beheshti, Tabebordbar and Benatallah (2020); Krzywinski and Cairo (2013); and Stolper et al., (2016) further emphasise the importance of telling a story with data, suggesting that skills associated with the Humanities, such as the analysis and construction of narratives, also play a critical role in data science and data analysis.

With respect to promoting knowledge and skills in the design and deployment of responsible AI in Africa, we recommend that educators integrate STEM with what Bourdeau and Wood (2019) call *humanistic STEM*, which they define as "blending the study of science, technology, engineering, and mathematics with interest in, and concern for, human affairs, welfare, values, or culture" (Bourdeau and Wood, 2019, p. 206). Such a recommendation clearly encapsulates the socio-technical view called for in this book that takes into consideration AI's social and technical facets, thus addressing what Tan (2021, p. 159) refers to as "the risk of opacity of technological function, and the hidden politics of technological systems". For Lina Markauskaite and her colleagues, explicit teaching of AI "includes the development of students and teachers' AI literacy [...] and humanistic thinking, such as ethics, philosophy, and historical ways of thinking" (Markauskaite et al., 2022, p. 13).

AI has the potential to positively transform labour markets in Africa but will also negatively disrupt them. While low levels of employment in Africa may be exacerbated by AI (Smith and Neupane, 2018, p. 77), the higher a person's qualifications, the smaller the risk that their job will be jeopardised by automation. Furthermore, qualifications increase an individual's chances of obtaining a permanent job, and qualified people also generally earn more than those who do not have secondary or tertiary qualifications. In addition to responding to unequal digital access, fostering responsible digital literacy (see Chapter 5), and helping young Africans focus on developing knowledge and skills in STEM disciplines that may also embrace a humanistic approach to science, technology, engineering, and mathematics, it is incumbent on African countries to ensure that complex tasks can be performed by workers and that laws and education are aligned with future demands in the context of the economy (Novitske, 2018). It is also essential that African countries tailor-make AI principles that speak to workers' socio-economic needs and address issues of justice and fairness in workspaces.

We turn now to a consideration of four prominent fields of enquiry, namely, machine learning, deep learning, text mining, and computer vision whose development and deployment have the potential for both good and ill on the continent.

Chapter 7

Machine learning, deep learning, text mining, and computer vision in Africa

Deployments and challenges

INTRODUCTION

AI has numerous definitions (for an overview, see Marr, 2018), but for the purpose of this chapter we agree that AI is an umbrella term that encapsulates the following basic idea: *machines (e.g., robots and computers) are expected to have human-like thought processes*. A classic example is the IBM supercomputer (Deep Blue) that outperformed the chess grandmaster Garry Kasparov in 1997. Another definitive example is Watson, a super-computer system that could compete against humans in real-time on the quiz show *Jeopardy!*. Examples such as these have sparked interest in AI from scholars and practitioners alike, an interest that has in turn been fuelled by the availability of huge datasets (Russell and Norvig, 2010, p. 27) that drive people's interest in smarter machines and artificial brains to analyse these datasets (Sharda, Delen and Turban, 2020, p. 79). As a field of research, AI has also advanced at a rapid pace over the past decade, given that researchers have made greater use of scientific methods in experimenting with and comparing approaches (Russell and Norvig, 2010, p. 30). As a result of the continued experimentation, our understanding of the theoretical basis for AI has improved significantly. This has resulted in an enormous ecosystem comprising major AI technologies, knowledge-based technologies, theories, tools, platforms, and mechanisms (Sharda et al., 2020, p. 53). These technologies and applications include autonomous vehicles, robo-advisors, augmented reality, smart homes, smart factories, smart cities, personal assistants, intelligent agents, voice recognition, machine learning, deep learning, neural networks, genetic algorithms, speech understanding, natural language processing, and computer vision (Sharda et al., 2020, p. 70). These are just selected examples of major elements of AI. For a more comprehensive list, see the 33 types of AI at https://simplicable.com/new/types-of-artificial-intelligence.

It is beyond the scope of this chapter to discuss all these elements and to consider the potential for good – and ill – that these fields hold for Africa.

DOI: 10.1201/9781003276135-7

For this reason, the focus falls specifically on four fields of enquiry that have become prominent in contributing towards addressing countries' socio-economic challenges, namely, machine learning, deep learning, text mining, and computer vision. These fields are visualised in Figures 7.1 and 7.2.

Figure 7.1 illustrates the major elements of AI, showing that deep learning, which is based on ANNs, is regarded as a separate field that has evolved from machine learning. At present, there are numerous models based on deep learning that have the capacity to outpace machine learning models in terms of performance (Janiesch, Zschech and Heinrich, 2021, p. 685). Figure 7.2 represents the various elements that fall under machine learning, text mining, and computer vision.

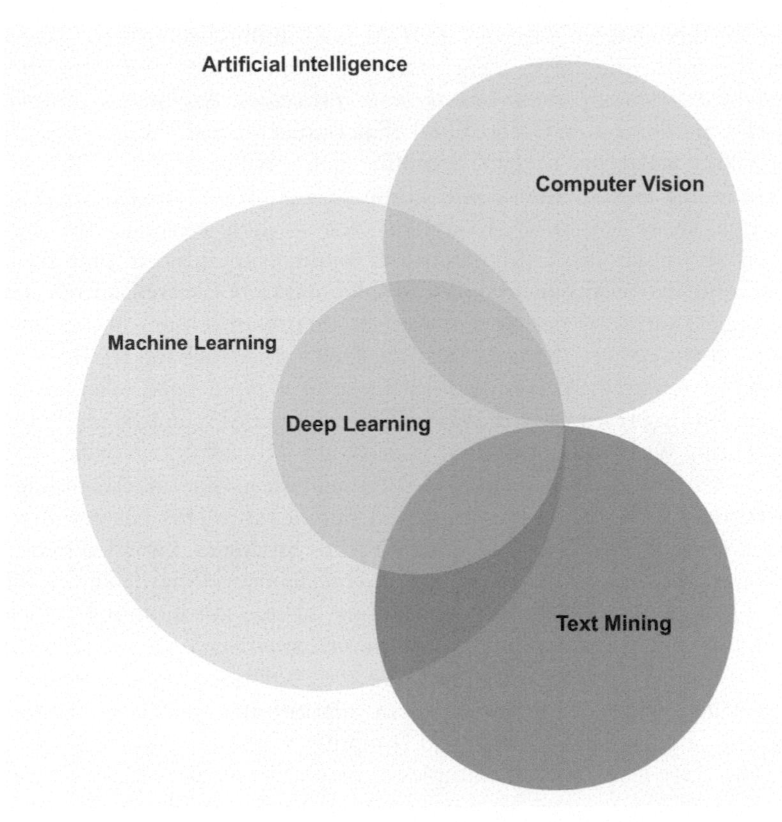

Figure 7.1 Major elements of AI.

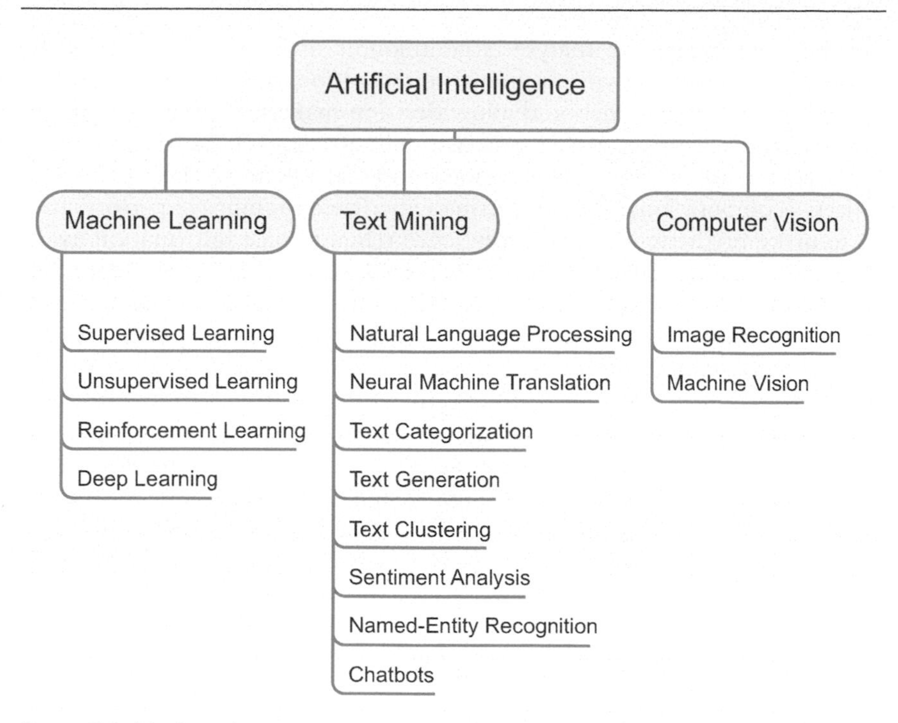

Figure 7.2 Machine learning, text mining, and computer vision.

MACHINE LEARNING

With the introduction of powerful computing hardware and the availability of voluminous historical datasets, a prominent branch of data science and business analytics referred to as predictive modelling has become increasingly prevalent. Predictive modelling – or predictive analytics – is forward-looking (i.e., with a focus on the future), using past events to anticipate the future (Eckerson, 2007). It therefore typically provides answers to the question *What might happen?* as opposed to the questions *Why did it happen?* and *What happened?*, which are traditionally associated with standard data analysis and reporting that employ business intelligence technologies (Eckerson, 2007). In simple terms, it allows decision-makers to "estimate" or "predict" what the future holds by learning from the past using historical data. Machine learning (ML) is a well-known example of a predictive analytics application and has evolved as a sub-field of AI (see Bon et al., 2022, p. 66) and even of computer science according to some perspectives since the second half of the twentieth century (Subasi, 2020, p. 93). Broadly speaking, ML employs self-learning algorithms to derive knowledge from data in order to make a prediction. In other words, instead of humans deriving rules and building models to analyse large amounts of data, ML captures the knowledge from data (without human intervention) to improve the performance of predictive models and allow decision-makers to make even better data-driven decisions (Raschka and Mirjalili, 2019, pp. 1–2).

According to Mohri, Rostamizadeh and Talwalkar (2012, p. 1), ML reflects "computational methods using experience to improve performance or to make predictions". Here experience refers to past information available to the learner, which typically takes the form of electronic data collected and made available for analysis. ML's myriad applications range from spam detection (text or document classification tasks), part-of-speech tagging and named-entity recognition (natural language processing) to speech recognition, face detection and image recognition (computer vision tasks), fraud detection, chess (games), medical diagnosis, search engines, and recommendation engines (Mohri et al., 2012, p. 2). This technology is of importance not only because it is making significant contributions to computer science research but also because it is playing an ever-increasing role in our daily lives. Today we employ ML in voice recognition software, web search engines such as Google and Yahoo, and even self-driving cars developed by Google. Most notable is the use of (deep) machine learning models to detect skin cancer with near-human level accuracy (Esteva et al., 2017).

DEEP LEARNING

It should be noted that deep learning is a sub-field of machine learning that employs representation learning (Goodfellow, Bengio and Courville, 2016, p. 9) to allow a computer to build complex concepts from simpler ones. For example, an image of a person is the combination of simpler concepts such as corners, contours, and edges (Goodfellow et al., 2016, p. 5). Deep learning employs a hierarchical and layered structure to represent the given input attributes (image of a person) utilising a nested, layered hierarchy of concept representations (i.e., corners, contours, and edges) when gaining experience (Subasi, 2020, p. 94). The most well-known example of a deep learning algorithm is the deep feedforward network or multiplayer perceptron (MLP), which in turn, is also a type of artificial neural network (Goodfellow et al., 2016, p. 168).

At this stage, Africa trails behind developed countries with respect to deploying ML to address a variety of socio-economic issues (cf. Elisa, 2018). It is fairly difficult to pinpoint the continent's specific challenges and risks associated with the design or deployment of ML, not only because of the informal nature of many ML activities in countries across the continent but also because of the paucity of information related to these activities in a variety of sectors (see Ly, 2021 in this regard). Nevertheless, it is still possible to identify obstacles and risks in major sectors such as agriculture,

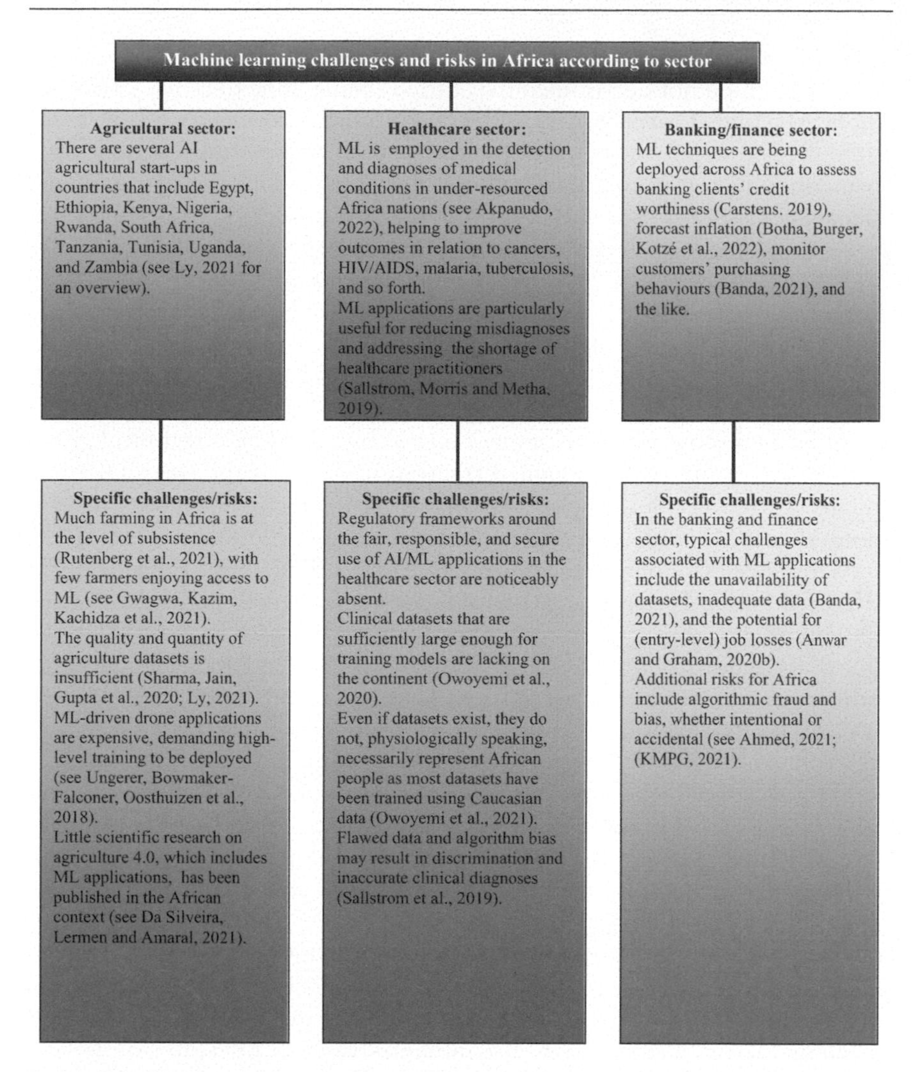

Figure 7.3 Challenges and risks associated with machine learning in Africa in three sectors.

healthcare, and banking and finance. Some of the main challenges and threats are illustrated in Figure 7.3.

While the obstacles and risks summarised in this figure are neither geographically nor topologically inclusive, they are nevertheless typical of African countries (and of many others in the Global South). Common obstacles to AI/ML deployments that are not unique to the continent (and which have been considered in previous chapters) encompass infrastructural deficits, technical issues, energy crises, and lack of funding.

Despite what some may view as Africa's somewhat haphazard approach to ML, it should not be assumed that little is being done to remedy the situation. Machine Learning Africa (https://machinelearningafrica.com/), for example, is a company based in South Africa that strives to create a platform on which various stakeholders that include innovators and end users may share their insight about ML innovations aimed at driving businesses and benefiting the public at large. Similarly, the Machine Intelligence Institute of Africa (https://miiafrica.org/) endeavours to develop an African AI ecosystem (comprised of key actors such as academics, the public sector, and businesses) that promotes AI, data science, and machine intelligence. The African Institute for Mathematical Sciences located in Rwanda offers an African Masters of Machine Intelligence aimed at fostering educational, scientific, and economic development (see Vernon, 2019). A technical grassroots organisation that has gained traction in Africa is the Deep Learning Indaba (https://deeplearningindaba.com/2022/), which endeavours to promote AI and ML with a view to achieving the goals of the so-called AI4SG or AI for Social Good movement (Tomašev et al., 2020, p. 1). The movement is geared towards creating AI-based projects aligned with the United Nation's sustainable development goals (SDGs) (Tomašev et al., 2020, p. 1). One of its major strengths lies in fostering AI development and education in indigenous communities (Chan et al., 2021), and a number of projects attest to this. These projects include the Snapshot Serengeti Challenge, which uses AI solutions to monitor the migration and behaviour of wildlife in Tanzania's Serengeti National Park (see Palmer et al., 2021), the development of natural language tools for low-resource languages in Africa (see Marivate, 2020), and the IBM Malaria Challenge (https://zindi.africa/competitions/ibm-malaria-challenge) established to find solutions to controlling the transmission of this febrile illness.

The AI4SG movement does, however, reflect an insidious side. In the context of public health in the United States, Holzmeyer (2021) interrogates AI4SG initiatives' ultimate consequences, offering the caveat that since their corporate actors include big tech companies such as Amazon, Apple, and Google who partner with organisations like the United Nations, what kind of social good in reality materialises may be questionable. Specifically, Holzmeyer (2021) avers that because AI4SG projects tend to frame problems as having technical solutions, they are inclined to "perpetuate incomplete, distorted models of social change that claim to be 'data-driven'" (Holzmeyer, 2021, p. 111). He goes on to argue that this unfortunately carries with it the risk of obfuscating these initiative's ethical consequences, thus widening social inequalities (Holzmeyer, 2021, p. 111).

Interestingly, and following Holzmeyer (2021), Moore (2019) calls for referring to "AI for Not Bad" rather than "AI for good", arguing that framing it in this way compels one to pay attention to avoiding the assumption that AI-enabled technologies must somehow be "good" (Moore, 2019, p. 5). Indeed, a number of scholars have critiqued AI for good or what has been referred to as "AI for SDGs" (Png, 2022, p. 1434), given that what "good" encompasses in practical terms has not been clearly conceptualised. Png (2022, p. 1434) references Green (2019) who argues that the definition of "social good" is vague and that there is in any case no universal consensus as to what it is supposed to entail. Further, Green (2019) notes that what is currently absent from computer science is the development of "a rigorous methodology for considering the relationship between algorithmic interventions and long-term social impact" (p. 2).

What is problematic about the assumption then is the disjunction between a perfect solution and what in reality occurs: "Pursuing social good without considering the long-term impacts can lead to great harm [...]: what may seem good in an immediate, narrow sense can be actively harmful in a broader sense" (Green, 2019, p. 3). AI's existing and potential ethical and societal costs for Africa have been detailed in previous chapters: its disruption and displacement of labour markets, its gendered and racialised dimensions, its surveillance of a country's citizens, its ecological impact, and the like. Chan et al. (2021) warn that if AI players do not form part of the Global South, then communities located in this region will reap few – if any – social and economic benefits from AI/ML initiatives. The crafting of Africa-inclusive AI policies and strategies that speak to the continent's unique contexts, the deployment of citizen science, and the integration of communities' indigenous knowledge systems into AI models have already been considered as possible solutions to addressing AI's dark side. What is encouraging is the establishment in 2020 of a four-year project referred to as AI4D Africa (Artificial Intelligence Development Africa) and initiated by the International Development Research Centre (IDRC) as well as the Swedish International Development Cooperation Agency (see Gwagwa et al., 2021). This initiative aims amongst other things, to encourage policy research with a view to advancing responsible AI and ML on the continent that aligns with local needs and speak to gender equity and sustainable development (IDRC, 2021). To date, a number of key partners have been selected by AI4D to assist African countries in their AI/ML policymaking endeavours. These partners include Niyel based in Senegal, the Centre for Intellectual Property and Information Technology (CIPIT) in Kenya, Research ICT Africa, and the African Observatory on Responsible AI (AORAI). Niyel (https://www.niyel.net/) is an advocacy consultant group that "seeks to establish a francophone network of researchers working on AI policy, build the capacity of research teams and policymakers, and support research teams in producing [...] research that will answer policymakers' needs" (IDRC, 2021, para. 9). The CIPIT (https://cipit.strathmore.edu/) makes use of knowledge and insights from computer science, political science, and law to garner and distribute information pertaining to intellectual property and information technology. Research ICT Africa is Southern Africa's think tank' on AI policymaking and encourages policies that are not centred around solutions to people's problems that are based on technological determinism (IDRC,

2021). Finally, the AORAI (https://www.africanobservatory.ai/) focuses on African AI ethics, AI and democracy, corporate accountability in the context of AI, and the like.

TEXT MINING

Text mining, also referred to as *text analytics* or *machine learning from text* (Aggarwal, 2018, p. xix), is an interdisciplinary field that draws on information retrieval, data mining, machine learning, deep learning, and natural language processing (NLP) to derive high-quality information from texts (Han, Kamber and Pei, 2012, p. 596; Sharda et al., 2020, p. 392). The field has become increasingly popular for researchers and practitioners in light of the explosion of text data generated on the Web and social media platforms (Aggarwal, 2018, p. 10). Examples of natural language texts include electronic news, digital libraries, email messages, blogs, and web pages. Some of the most widespread applications for text mining include machine translation, text categorisation, text summarisation, text clustering, sentiment analysis, entity-relation modelling, question answering, and chatbots (Han et al., 2012, p. 597; Sharda et al., 2020, p. 394).

Since computers cannot understand texts in their native format, simplified representation approaches are required before machine learning models can be trained using texts. Texts, such as a sentence, a paragraph, or a complete document, can be represented as *bag-of-words* where the set of words is converted into a sparse multidimensional representation (Aggarwal, 2018, p. xx). Neither the ordering of the words nor the grammar matters. The terms (or *universe of words*) will correspond to the dimensions (or *features*) in this presentation, which can then be used for applications such as machine translation, text classification, topic modelling, and recommender systems. Text can also be presented as a *set of sequences*, where sentences in a document are extracted as strings or sequences (Aggarwal, 2018, p. xxi). In this approach, the ordering of the words in the representation is important, even though it is only related to a sentence or paragraph. This approach is required by applications that rely on the semantic interpretation of the document content such as *language modelling* and *natural language processing* (part-of-speech tagging, named-entity recognition), though the latter is often treated as a separate research discipline (Aggarwal, 2018, p. xxi). Thereafter, sequence-centric models will employ this approach to transform the sequential representation of text (i.e., *text as a sequence*) to a multidimensional representation. A neural network method such as *word2vec* (Mikolov et al., 2013) is a very useful representation learning method to transform documents and leverage sequential language models to create multidimensional features which are referred to as *multidimensional embeddings* (Aggarwal, 2018, p. 9). For example, if we consider the word analogy "king is to queen as man is to woman", then we can express the multidimensional embeddings *function* of the terms "*king*", "*queen*", "*man*", and "*woman*" as follows:

$$f(king) - f(queen) \approx f(man) - f(woman)$$

It should be noted that the above function can only be created with the use of sequence information because of the semantic nature of the words and in the end, result in a semantically knowledge embeddings (Aggarwal, 2018, p. 301). Machine learning solutions can thereafter be trained on the embeddings and used for natural language processing applications (Aggarwal, 2018, pp. 30–302). For example, in neural machine translation (NMT), an artificial neural network (ANN) is used to predict the likelihood of a sequence of words, which is typically an entire sentence using vector representations such as embeddings (or "continuous space representations") for words.

Research studies in text mining and NLP from Global South countries are not well represented at international research conferences on language technologies. In a 2019 article, Andrew Caines from the Department of Computer Science and Technology at the University of Cambridge reported that a mere five full-text papers from Africa formed part of five 2018 NLP conferences in the ACL [Association for Computational Linguistics] Anthology, despite the fact that more than 2000 languages are spoken in Africa (see Eberhard, Simons and Fennig, 2019). This is in marked contrast to 1114 full-text papers from North America, 826 from Asia, and 641 from Europe (Caines, 2019).

On the other hand, a number of initiatives are afoot to rectify this underrepresentation on the continent. Google, for example, recently contributed to the language technology field by adding several African languages to Google Translate, which include Bambara, Ewe, Tsonga Luganda, and Twi (https://blog.google/products/translate/24-new-languages/). An exciting recent event, and still ongoing at the time of writing this book, is the "Google NLP Hack Series: Swahili Social Media Sentiment Analysis Challenge" (https://zindi.africa/competitions/swahili-social-media-sentiment-analysis-challenge) established to encourage text mining research from East African countries (Malawi, Rwanda, Uganda, Kenya, and Tanzania) to determine if the sentiment of a tweet is positive, negative, or neutral. Specifically, Swahili text sentences were collected from Twitter and annotated to provide a corpus that can be used to analyse social conversations online and determine deeper context. Masakhane ("Let us build together" in isiZulu) is a grassroots organisation founded to improve NLP participation in African communities (https://www.masakhane.io/). The research effort initially focused on machine translation of African languages (https://africanlp.masakhane.io/) but soon expanded to include language models for African languages, named entity recognition (NER), and the annotation of high-quality text and speech datasets for low-resource East African languages (see Orife et al., 2020).

What is sobering are the reasons Abbott and Martinus (2019) put forward as to why NLP has not been widely adopted in Africa. These include low levels of confidence among local communities that their indigenous languages (as opposed to hegemonic ones) may function as primary modes of communication, low availability of – and difficulty in – obtaining resources for African languages, and the reluctance of researchers to make their data publicly available. Employing NMT techniques to train models to translate English into a number of South Africa's official languages, Abbott and Martinus (2019) suggest that the source codes and data for projects should be discoverable as well as reproducible and that benchmarks should be made available so that other researchers may expand on them and compare machine translation techniques to others.

COMPUTER VISION

Computer vision (CV) is a sub-field of artificial intelligence that enables computers to extract high-level information from digital images and videos. If AI enables computers to "think", then computer vision enables them to "see" and "observe"[1]. The goal of computer vision is to understand the content of digital images or videos, thus developing methods to reproduce the capability of human vision (Forsyth and Ponce, 2003, p. xvii). It is therefore easy to assume that computer vision can be equated to human vision. However, it employs data and algorithms instead of retinas, optic nerves, and the visual cortex to understand an image. In the context of computer vision, "understanding" means transforming the visual image (i.e., the input of the retina) into descriptions of the world that would make sense. Computer vision should not be confused with image processing, which is the process of creating a new image from an existing image. Current computer vision applications rely on machine learning tools to process images, particularly deep learning models such as a convolutional neural network (CNN) to extract information from an image (Goodfellow et al., 2016, p. 254). The information can include 3D models, camera position, object detection, and the like (Solem, 2012, p. ix). The most common applications of computer vision include scene reconstruction, object detection, video tracking, object recognition, motion estimation, 3D scene modelling, and image restoration.

One of the most well-known applications is medical computer vision (or medical image processing), where information for image data is extracted to diagnose a disease. Military applications are also widespread and include the detection of enemy soldiers (or vehicles) and missile guiding systems. Autonomous vehicles such as NASA's *Curiosity* make use of computer vision to navigate the landscape of Mars. In terms of object recognition, a recent example is Google Lens app (https://lens.google/), a mobile device service launched in 2018 using deep learning artificial neural network algorithms (and other AI techniques) to identify images using a mobile phone (Sharda et al., 2020, p. 354).

At the Council for Scientific and Industrial Research (CSIR) based in South Africa, CSIR (2022), researchers have designed algorithms to detect and track people for use in a human-following robot. The algorithms are also able to recognise human hand gestures to that commands can be given to a robot. Computer vision and machine learning are also used in an ongoing project called Zamba, which means "forest" in *Lingala*, a language used in the Democratic Republic of the Congo. The project is aimed at automatically detecting and classifying animals in camera trap videos and is driven by software developed by DrivenData (2021). In a recent research report by Mudongo (2021), computer vision work conducted in Africa was put under the spotlight. This work includes the closed circuit television (CCTV) surveillance network in Botswana, the CCTV network in the City of Cape Town, and the CCTV network in the City of Johannesburg. The report concludes that existing legal and governance frameworks for the use of computer vision for automated surveillance are inadequate in the three case studies referred to.

Machine learning, deep learning, text mining, and computer vision are just some of the technologies that are currently being employed in Africa to improve conditions in areas such as agriculture, business and finance, healthcare, and wildlife conservation. However, these technologies can also be used for ill if not carefully regulated and monitored.

NOTE

1 https://www.ibm.com/za-en/topics/computer-vision

Postface

What cannot be refuted is the substantial contribution AI-enabled technologies can make in solving some of Africa's most pressing socio-economic problems that are partially exacerbated as a result of colonialism, which remains alive and well across many parts of the continent. As they pertain to the fourth industrial revolution on the continent, colonial legacies are numerous and include cyber colonisation, data ownership in the hands of a few technological oligarchs, exploitation of workers in the gig economy, the formation of digital dictatorships, ongoing digital divides, and the like. What this in turn means is that what countries in Africa do not need is a design or deployment of AI that takes into account only technocratic and/or West-centric solutions to their unique challenges. What is called for is a humanistic approach to AI – to an acknowledgement of *all* AI-stakeholders-in-the-loop so that plurality and pluriversality are given free rein in AI ecosystems that in turn must be established along responsible and ethical lines. However, such an approach cannot be followed without the prudent and thoroughly researched drafting of Africa-inclusive AI policies. Policy responses to AI are slowly but steadily materialising, and what we advocate is that these policies incorporate a relational ethics of care that explicitly speaks to how Africa's disparities – such as those related to gender, race, and labour – are entwined in AI ecosystems. Such an ethics of care underscores the fact that in the process of creating policies, tenets such as inclusion, diversity, pluralism, and advocacy should be transformed into *measurable, evidence-based observations* in all sectors of society that are touched by AI. In this way, policymaking will avoid the decontextualisation of AI, which risks disregarding the realities of people in Africa who may, on the one hand, be vulnerable to AI's threats and, on the other, be entirely excluded from this technology's benefits. This humanistic approach embraces the notion that while technology may be beneficial, it remains in the hands of a few and therefore carries with it the threat of perpetuating digital coloniality (Mutsvairo, Ragnedda and Orgeret, 2021, p. 296).

DOI: 10.1201/9781003276135-8

References

Abayomi, O.K., Adenekan, F.N., Abayomi, A.O., Ajayi, T.A. and Aderonke, A.O. (2021) 'Awareness and perception of the artificial intelligence in the management of University Libraries in Nigeria', *Journal of Interlibrary Loan, Document Delivery & Electronic Reserve*, 29(1–2), pp. 13–28. doi:10.1080/1072303X.2021.1918602.

Abbott, J. and Martinus, L. (2019) 'Benchmarking neural machine translation for Southern African languages', in Axelrod, A., Yang, D., Cunha, R., Shaikh, S. and Waseem, Z. (eds.) *Proceedings of the 2019 Workshop on Widening NLP*. Florence, Italy: Association for Computational Linguistics, pp. 98–101.

Absar, S. (2020) 'The troubling return of scientific racism', *Varsity*, 20 June. Available at: https://www.varsity.co.uk/science/19479.

Acemoglu, D. and Restrepo, P. (2017) *Robots and Jobs: Evidence from US Labor Markets*. Cambridge, MA: National Bureau of Economic Research. doi:10.3386/w23285.

Adam, T. (2019) 'Digital neocolonialism and massive open online courses (MOOCs): colonial pasts and neoliberal futures', *Learning, Media and Technology*, 44(3), pp. 365–380. doi:10.1080/17439884.2019.1640740.

Adams, N.R. (2019) 'Decolonising the origins of artificial intelligence', *Medium*, 2 October. Available at: https://becominghuman.ai/decolonising-the-origins-of-artificial-intelligence-3af4bb419816.

Adams, R. (2019) 'The Fourth Industrial Revolution risks leaving women behind', *World Economic Forum* in Collaboration with *The Conversation*, 7 August. Available at: https://theconversation.com/the-fourth-industrial-revolution-risks-leaving-women-behind-121216.

Adams, R. (2021) 'Can artificial intelligence be decolonised?', *Interdisciplinary Science Reviews*, 46(1–2), pp. 176–197. doi:10.1080/03080188.2020.1840225.

Adeniran, A.P., Ekeruche, M.A., Onyekwena, C. and Obiakor, T. (2021) 'Estimating the economic impact of Chinese BRI investment in Africa', Special Report, June 2021. *South African Institute of International Affairs*. Available at: https://media.africaportal.org/documents/Special-Report-adeniran-ekeruche-onyekwena-obiakor.pdf.

Adeyoju, A. (2018) 'Artificial intelligence and the future of law practice in Africa', *SSRN*, 15 December. Available at: https://ssrn.com/abstract=3301937. doi:10.2139/ssrn.3301937.

African Development Bank/AfDB. (2018) *African Economic Outlook 2018*. Abidjan, Cote d'Ivoire: AfDB.

African Union. (2020) 'The digital transformation strategy for Africa (2020–2030)', *African Union*. Addis Ababa, Ethiopia. Available at: https://au.int/sites/default/files/documents/38507-doc-dts-english.pdf.

AFP. (2018a) 'AI better at finding skin cancer than doctors: study', *Daily Maverick*, 29 May. Available at: https://www.dailymaverick.co.za/article/2018-05-29-ai-better-at-finding-skin-cancer-than-doctors-study/.

AFP. (2018b) 'Self-navigating AI learns to take shortcuts: study', *The Citizen*, 9 May. Available at: https://www.citizen.co.za/news/news-world/1920598/self-navigating-ai-learns-to-take-shortcuts-study/.

AFP. (2019) 'Artificial intelligence trained to identify lung cancer', *The Citizen*, 22 May. Available at: https://www.citizen.co.za/lifestyle/health/2133659/artificial-intelligence-trained-to-identify-lung-cancer/.

Aggarwal, C.C. (2018) *Machine Learning for Text*. Berlin/Heidelberg, Germany: Springer.

Agyekum, A. (2021) 'Government urged to act swiftly to prevent "killer robots" development', *Ghana News Agency*, 27. Available at: https://www.gna.org.gh/1.21160683.

Ahlborg, H., Ruiz-Mercado, I., Molander, S. and Masera, O. (2019) 'Bringing technology into social-ecological systems research: motivations for a socio-technical-ecological systems approach', *Sustainability*, 11(7), pp. 1–23. doi:10.3390/su11072009.

Ahmed, S. (2020) 'Africa needs sustainable digitalisation: the link between tech, inequality and the environment', *Research ICT Africa*, 14 October. Available at: https://researchictafrica.net/2020/10/14/africa-needs-sustainable-digitalisation-for-post-covid-19-future-resilience/.

Ahmed, S. (2021) 'A gender perspective on the use of artificial intelligence in the African Fintech ecosystem: case studies from South Africa, Kenya, Nigeria and Ghana'. Paper presented at the *International Telecommunications Society (ITS) 23rd Biennial Conference – Digital Societies and Industrial Transformations: Policies, Markets, and Technologies in a Post-Covid World*, June 21–23, 2021.

Akanbi, A.K. and Masinde, M. (2018) 'Towards the development of a rule-based drought early warning expert systems using indigenous knowledge', *2018 International Conference on Advances in Big Data, Computing and Data Communication Systems (icABCD)*. IEEE, pp. 1–8. doi:10.1109/ICABCD.2018.8465465.

Akpanudo, S. (2022) 'Application of artificial intelligence systems to improve health-care delivery in Africa', *Primary Health Care: Open Access*, 12(1), pp. 1–4.

Alami, H., Rivard, L., Lehoux, P., Hoffman, S.J., Cadeddu, S.B.M., Savoldelli, M., Samri, M.A., Ahmed, M.A.A., Fleet, R. and Fortin, J.P. (2020) 'Artificial intelligence in health care: laying the foundation for responsible, sustainable, and inclusive innovation in low-and middle-income countries', *Globalization and Health*, 16(1), pp. 1–6. doi:10.1186/s12992-020-00584-1.

Alamu, F.O., Aworinde, H.O. and Isharufe, W.I. (2013) 'A comparative study on IFA divination and computer science', *International Journal of Innovative Technology and Research*, 6(1), pp. 524–528.

Alupo, C.D., Omeiza, D. and Vernon, D. (2022) 'Realizing the potential of AI in Africa: it all turns on trust', in Aldinhas Ferreira, M.I. and Tokhi, O. (eds.) *Towards Trustworthy Artificial Intelligence Systems*. Cham, Switzerland, Springer, pp. 179–192.

Amadasun, K.N., Short, M., Shankar-Priya, R. and Crosbie, T. (2021) 'Transitioning to society 5.0 in Africa: tools to support ICT infrastructure sharing', *Data*, 6(7), pp. 1–18. doi:10.3390/data6070069.

Annas, J. (1993) *The Morality of Happiness*. New York, NY: Oxford University Press.

Anthonysamy, Lilian (2020) 'Digital literacy deficiencies in digital learning among undergraduates', in Noviaristanti, S., Hanafi, H. and Trihanondo, D. (eds.) *Understanding Digital Industry*. London, UK: Routledge, pp. 133–136. doi:10.1201/9780367814557.

Antwi, W.K., Akudjedu, T.N. and Botwe, B.O. (2021) 'Artificial intelligence in medical imaging practice in Africa: a qualitative content analysis study of radiographers perspectives' *Insights Imaging*, 12(80), pp. 1–9. doi:10.1186/s13244-021-01028-z.

Anwar, M.A. and Graham, M. (2020a) 'Hidden transcripts of the gig economy: labour agency and the new art of resistance among African gig workers', *Environment and Planning A: Economy and Space*, 52(7), pp. 1269–1291. doi:10.1177/0308518X19894584.

Anwar, M.A. and Graham, M. (2020b) 'Digital labour at economic margins: African workers and the global information economy', *Review of African Political Economy*, 47(163), pp. 95–105. doi:10.1080/03056244.2020.1728243.

Appiah, E.K., Arko-Achemfuor, A. and Adeyeye, O.P. (2018) 'Appreciation of diversity and inclusion in sub-Sahara Africa: the socioeconomic implications', *Cogent Social Sciences*, 4(1), pp. 1–12. doi:10.1080/23311886.2018.1521058.

Armitage, A. (2018) 'Is HRD in need of an ethics of care?', *Human Resource Development International*, 21(3), pp. 212–231. doi:10.1080/13678868.2017.1366176.

Arnold, T., Kasenberg, D. and Scheutz, M. (2017) 'Value alignment or misalignment: what will keep systems accountable?', in *Workshops at the Thirty-First AAAI Conference on Artificial Intelligence*. USA: AAAI.

Arntz, M., Gregory, T. and Zierahn, U. (2016) *The Risk of Automation for Jobs in OECD Countries: A Comparative Analysis*, Working Paper No. 189. Paris, France: OECD Publishing. doi:10.1787/5jlz9h56dvq7-en.

Arrighi, G. (2010) *The Long Twentieth Century: Money, Power, and the Origins of Our Times*. London: New York, NY: Verso.

Asaro, P.M. (2019) 'AI ethics in predictive policing: from models of threat to an ethics of care', *IEEE Technology and Society Magazine*, 38(2), pp. 40–53.

Atkinson, R.D. (2016) '"It's Going to Kill Us!" and other myths about the future of artificial intelligence', *Information Technology & Innovation Foundation*, June 2016. Available at: https://www2.itif.org/2016-myths-machine-learning.pdf?_ga=2.17408889.1347365343.1639036002-1151542750.1639036002.

Auriacombe, C. and Van der Walt, G. (2021) 'Fundamental policy challenges influencing sustainable development in Africa', *Africa's Public Service Delivery and Performance Review*, 9(1), pp. 1–8. doi:10.4102/apsdpr.v9i1.381.

Autor, D.H. and Dorn, D. (2013) 'The growth of low-skill service jobs and the polarization of the US labor market', *American Economic Review*, 103(5), pp. 1553–1597. doi:10.1257/aer.103.5.1553.

Bakhshi, H., Downing, J.M., Osborne, M.A. and Schndier, P. (2017) *The Future of Skills*. London, UK: Pearson and Nesta.

Balayn, A. and Gürses, S. (2021) *Beyond Debiasing: Regulating AI and its Inequalities*. Brussels, Belgium: European Digital Rights (EDRi). Available at: https://edri.org/wp-content/uploads/2021/09/EDRi_Beyond-Debiasing-Report_Online.pdf.

Banda, M. (2021) 'Machine Learning Garners Impetus in Africa', *Intelligent CIO*, 6 July. Available at: https://www.intelligentcio.com/africa/2021/07/06/machine-learning-garners-impetus-in-africa/.

Barbat, M.T.G. (2018) 'Foreword', in Metzinger, T., Bentley, P.J., Häggström, O. and Brundage, M. (eds.) *Should We Fear Artificial Intelligence?*. Brussels: EPRS European Parliamentary Research Centre, p. 4.

Bardzell, S. (2010) 'Feminist HCI: taking stock and outlining an agenda for design', in Konstan, J.A., Chi, E. and Höök, K. (eds.) *CHI '10: Proceedings of the SIGCHI Conference on Human Factors in Computing Systems*. New York, NY: ACM Press, pp. 1301–1310.

Bartneck, C. Yogeeswaran, K., Ser, Q.M., Woodward, G., Sparrow, R., Wang, S. and Eyssel, F. (2018) 'Robots and racism', in Kanda, T., Sabanovic, S., Hoffman, G. and Tapus, A. (eds.) *Proceedings of the 2018 ACM/IEEE International Conference on Human-robot Interaction*. Chicago, IL: ACM, pp. 196–204. doi:10.1145/3171221.3171260.

Barton, A.M.C. (2012) 'Citizen(s') science: a response to "The Future Of Citizen Science"', *Democracy & Education*, 20(2), pp. 1–4.

Basu, S. and Weil, D.N. (1998) 'Appropriate technology and growth', *Quarterly Journal of Economics*, 113(4), pp. 1025–1054.

Behari-Leak, K. (2019) 'Decolonial turns, postcolonial shifts, and cultural connections: are we there yet?', *English Academy Review*, 36(1), pp. 58–68. doi:10.1080/10131752.2019.1579881.

Beheshti, A., Tabebordbar, A. and Benatallah, B. (2020) 'iStory: intelligent storytelling with social data', in Seghrouchni, A.E.F., Sukthankar, G., Liu, T.-Y. and Van Steen, M. (eds.) *Companion Proceedings of the Web Conference 2020. WWW '20: The Web Conference 2020*. New York, NY: ACM, pp. 253–256. doi:10.1145/3366424.3383553.

Behymer, K.J. and Flach, J.M. (2016) 'From autonomous systems to sociotechnical systems: designing effective collaborations', *She Ji: The Journal of Design, Economics, and Innovation*, 2(2), pp. 105–114. doi:10.1016/j.sheji.2016.09.001.

Bendig, E., Erb, B., Schulze-Thuesing, L. and Baumeister, H. (2019) 'The next generation: chatbots in clinical psychology and psychotherapy to foster mental health – a scoping review', *Verhaltenstherapie*: 1–13. doi:10.1159/000501812.

Bentley, P. (2018) 'The three laws of artificial intelligence: dispelling common myths', in Metzinger, T., Bentley, P.J., Häggström, O. and Brundage, M. (eds.) *Should We Fear Artificial Intelligence?*. Brussels: EPRS European Parliamentary Research Centre, pp. 6–12.

Benyera, E. (2021) *The Fourth Industrial Revolution and the Recolonisation of Africa: The Coloniality of Data*. New York, NY: Routledge.

Berkes, F. and Berkes, M.K. (2009) 'Ecological complexity, fuzzy logic, and holism in indigenous knowledge', *Futures*, 41(1), pp.6–12. doi:10.1016/j.futures.2008.07.003.

Bertani, F., Raberto, M. and Teglio, A. (2020) 'The productivity and unemployment effects of the digital transformation: an empirical and modelling assessment', *Review of Evolutionary Political Economy*, 1(3), pp. 329–355. doi:10.1007/s43253-020-00022-3.

Besaw, C. and Filitz, J. (2019) 'Artificial intelligence in Africa is a double-edged sword', *Our World*, 16 January. Available at: https://ourworld.unu.edu/en/ai-in-africa-is-a-double-edged-sword.

Bhatasara, S. and Chirimambowa, T.C. (2018) 'The gender and labour question in the future of work discourses in Southern Africa', *BUWA!*, (9), pp. 23–28.

Binet, A. (1905) 'New methods for the diagnosis of the intellectual level of subnormals', *L'Anne´e Psychologique*, 12, pp. 191–244.

Biney, A. (2014) 'The historical discourse on African humanism: interrogating the paradoxes', in Praeg, L. and Magadla, S. (eds.) *Ubuntu: Curating the Archive*. Pietermaritzburg, South Africa: University of KwaZulu-Natal Press, pp. 27–53.

Birhane, A. (2020a) 'Algorithmic colonization of AI', *Scripted*, 17(2), pp. 389–409. doi:10.2966/scrip.170220.389.

Birhane, A. (2020b) 'Aglorithmic colonisation of Africa', *The Elephant*, 21 August. Available at: https://www.theelephant.info/long-reads/2020/08/21/algorithmic-colonisation-of-africa/?print=pdf.

Birhane, A. (2021) 'Algorithmic injustice: a relational ethics approach', *Patterns*, 2(2), pp. 1–9. doi:10.1016/j.patter.2021.100205.

Bissessur, J., Arabikhan, F. and Bednar, P. (2020) 'The illusion of routine as an indicator for job automation with artificial intelligence', in Lazazzara, A., Ricciardi, F. and Za, S. (eds.) *Exploring Digital Ecosystems: Organizational and Human Challenges*. Cham, Switzerland: Springer International Publishing, pp. 407–416. doi:10.1007/978-3-030-23665-6_29.

Black, E. and Richmond, R. (2019) 'Improving early detection of breast cancer in sub-Saharan Africa: why mammography may not be the way forward', *Globalization and Health*, 15(1), pp. 1–11. doi:10.1186/s12992-018-0446-6.

Boch, A., Lucaj, L. and Corrigan, C. (2021) 'A robotic new hope: opportunities, challenges, and ethical considerations of social robots', *Research Brief*. Technical University of Munich: Munich Center for Technology in Society and Institute for Ethics in Artificial Intelligence (IEAI). Available at: https://ieai.mcts.tum.de/wp-content/uploads/2021/05/ResearchBrief_April2021_SocialRobots_FinalV2.pdf.

Bon, A., Dittoh, F., Lô, G., Pini, M., Bwana, R., WaiShiang, C., Kulathuramaiyer, N. and Baart, A. (2022) 'Decolonizing technology and society: a perspective from the global south', in Werthner, H., Prem, E., Lee, E.A. and Ghezzi, C. (eds.) *Perspectives on Digital Humanism*. Cham, Switzerland, pp. 61–70. doi:10.1007/978-3-030-86144-5.

Borenstein, J., Herkert, J.R. and Miller, K.W. (2019) 'Self-driving cars and engineering ethics: the need for system level analysis', *Science and Engineering Ethics*, 25(2), pp. 383–398. doi:10.1007/s11948-017-0006-0.

Borgonovi, F., Centurelli, R., Dernis, H., Grundk, E.R., Horvát, P., Jamet, S., Keese, M., Liebender, A.-S., Marcolin, L., Rosenfeld, D. and Squicciarini, M. (2018) *Bridging the Digital Gender Divide*. Australia: OECD Directorate for Science, Technology and Innovation (STI), Directorate for Education and Skills (EDU) & Directorate for Employment, Labour and Social Affairs (ELS).

Botha, B. Burger, R., Kotzé, K., Rankin, N. and Steenkamp, D. (2022) 'Big data forecasting of South African inflation', *South African Reserve Bank Working Paper Series WP/22/01*. South Africa: South African Reserve Bank. Available at: https://www.resbank.co.za/content/dam/sarb/publications/working-papers/2022/WP%202201.pdf.

Bourdeau, D.T. and Wood, B.L. (2019) 'What is humanistic STEM and why do we need it?', *Journal of Humanistic Mathematics*, 9(1), pp. 205–216. doi:10.5642/jhummath.202101.04.

Botswana Guardian reporter. (2021) 'Tech innovation enlisted in GBV war', *Botswana Guardian*, 30 July. Available at: https://www.pressreader.com/botswana/botswana-guardian/20210730/281685437881500.

Bourne, C. (2019) 'AI cheerleaders: public relations, neoliberalism and artificial intelligence', *Public Relations Inquiry*, 8(2), pp. 109–125. doi:10.1177/2046147X19835250.

Boykoff, M.T. and Boykoff, J.M. (2004) 'Balance as bias: global warming and the US prestige press', *Global Environmental Change*, 14(2), pp. 125–136. doi:10.1016/j.gloenvcha.2003.10.001.

Bozkaya, I. (2021) 'The importance of eudaimonia for action-guiding virtue ethics', *The Journal of Value Inquiry*, pp. 1–16. doi: 10.1007/s10790-021-09816-y.

Bradshaw, S., Bailey, H. and Howard, P.N. (2021) *Industrialized Disinformation: 2020 Global Inventory of Organised Social Media Manipulation, Working Paper 2021.1*. Oxford, UK: Project on Computational Propaganda. Available at: https://demtech.oii.ox.ac.uk/wp-content/uploads/sites/127/2021/01/CyberTroop-Report20-FINALv.3.pdf.

Brigham, C.C. (1923) *A Study of American Intelligence*. Princeton, NJ: Princeton University Press.

Bringsjord, S. (2013) *What Robots Can and Can't Be* (vol. 12). USA: Springer Science & Business Media. doi:10.1007/978-94-011-2464-5.

Brokensha, S.I. (2020) 'Friend or foe? How online news outlets in South Africa frame artificial intelligence', *Ensovoort*, 41(7), p. 2. Available at: http://ensovoort.com/friend-orfoe-how-online-news-outlets-in-south-africa-frame-artificial-intelligence/.

Brokensha, S.I. and Conradie, T. (2021) 'Killer robots, humanoid companions, and superintelligent machines: the anthropomorphism of AI in South African news articles', *Ensovoort*, 42(6), p. 3. Available at: http://ensovoort.com/killer-robots-humanoidcompanions-and-super-intelligent-machines-the-anthropomorphism-of-ai-in-southafrican-news-articles/.

Brożek, B. and Janik, B. (2019) 'Can artificial intelligences be moral agents?', *New Ideas in Psychology*, 54, pp. 101–106. doi:10.1016/j.newideapsych.2018.12.002.

Brown, W. (2006) 'American nightmare: neoliberalism, neoconservatism, and de-democratization', *Political Theory*, 34(6), pp. 690–714.

Bruchac, M. (2014) 'Indigenous knowledge and traditional knowledge', in Smith, C. (ed.) *Encyclopedia of Global Archaeology*. New York, NY: Springer, pp. 3814–3824. doi:10.1007/978-1-4419-0465-2.

Bruun, E.P.G. and Duka, A. (2018) 'Artificial intelligence, jobs and the future of work: racing with the machines', *Basic Income Studies*, 13(2). doi:10.1515/bis-2018-0018.

Brynjolfsson, E. and Mitchell, T. (2017) 'What can machine learning do? Workforce implications', *Science*, 358(6370), pp. 1530–1534. doi:10.1126/science.aap8062.

Burton, E., Goldsmith, J., Koenig, S., Kuipers, B., Mattei, N. and Walsh, T. (2017) 'Ethical considerations in artificial intelligence courses', *AI Magazine*, 38(2), pp. 22–34.

Burton, K. (2019) 'Artificial intelligence to predict water scarcity conflicts', *Geographical*. Available at: https://geographical.co.uk/places/wetlands/item/3324-artificial-intelligence-used-to-help-predict-potential-conflicts-related-to-water-scarcity (Accessed: 13 August 2019).

Butcher, N., Wilson-Strydom, M. and Baijnath, M. (2021) 'Artificial intelligence capacity in sub-Saharan Africa: compendium report', *International Development*

Research Centre. Available at: https://africa.ai4d.ai/wp-content/uploads/2021/03/AI4DReport%E2%80%94AI-in-SSA.pdf.

Bynum, T.W. (2006) 'Flourishing ethics', *Ethics and Information Technology*, 8(4), pp. 157–173. doi:10.1007/s10676-006-9107-1.

Caboz, J. (2021) 'We "hired" an AI writing assistant to create an article for us – our jobs are safe for now', *Business Insider SA*. Available at: https://www.businessinsider.co.za/we-hired-an-ai-copywriting-assistant-to-create-an-article-for-us-our-jobs-are-safe-for-now-2021-10 (Accessed: 17 October 2021).

Cabrales, A., Hernández, P. and Sánchez, A. (2020) 'Robots, labor markets, and universal basic income', *Humanities and Social Sciences Communications*, 7(1), p. 185. doi:10.1057/s41599-020-00676-8.

Cafaro, P. (2015) 'Environmental virtue ethics', in Besser-Jones, L. and Slote, M. (eds.) *The Routledge Companion to Virtue Ethics*. New York, NY: Routledge, pp. 451–468.

Caines, A. (2019) 'The geographic diversity of NLP conferences', *Marek Rei*, 4 October. Available: http://www.marekrei.com/blog/geographic-diversity-of-nlp-conferences/.

Campolo, A. and Crawford, K. (2020) 'Enchanted determinism: power without responsibility in artificial intelligence', *Engaging Science, Technology, and Society*, 6, pp.1–19. doi:10.17351/ests2020.277.

Canales, C., Lee, C. and Cannesson, M. (2020) 'Science without conscience is but the ruin of the soul: the ethics of big data and artificial intelligence in perioperative medicine', *Anesthesia and Analgesia*, 130(5), pp. 1234–1243. doi:10.1213/ANE.0000000000004728.

Candelon, F., Bedraoui, E. and Maher, H. (2021) 'Developing an artificial intelligence for Africa strategy', *OECD Development Matters*, 9 February. Available at: https://oecd-development-matters.org/2021/02/09/developing-an-artificial-intelligence-for-africa-strategy/.

Carman, M. and Rosman, B. (2021) 'Applying a principle of explicability to AI research in Africa': should we do it?', *Ethics and Information Technology*, 23(2), pp. 107–117. doi:10.1007/s10676-020-09534-2.

Carstens, A. (2019) 'Big tech in finance and new challenges for public policy', *SUERF Policy Note*, 54, pp. 1–12.

Casilli, A.A. (2021) 'Waiting for robots: the ever-elusive myth of automation and the global exploitation of digital labor', *Sociologias*, 23(57), pp. 112–133. doi:10.1590/15174522-114092.

Castelvecchi, D. (2016) 'Can we open the black box of AI?', *Nature*, 538(7623), pp. 20–23. doi:10.1038/538020a.

Caswell, I. and Liang, B. (2020) 'Recent advances in google translate', *Google AI Blog*, 8 June. Available at: https://ai.googleblog.com/2020/06/recent-advances-in-google-translate.html.

Cave, S. (2020) 'The problem with intelligence: its value-laden history and the future of AI', in Markham, A. Powles, J., Walsh, T. and Washington, A.L. (eds.) *Proceedings of the AAAI/ACM Conference on AI, Ethics, and Society*. New York, NY: ACM, pp. 29–35. doi:10.1145/3375627.3375813.

Cave, S., Craig, C., Dihal, K., Dillon, S., Montgomery, J., Singler, B. and Taylor, L. (2018) *Portrayals and Perceptions of AI and Why They Matter*. London, UK: The Royal Society. Available at: https://royalsociety.org/~/media/policy/projects/ai-narratives/AI-narratives-workshop-findings.pdf.

Cave, S. and Dihal, K. (2020) 'The whiteness of AI', *Philosophy & Technology*, pp. 1–19. doi:10.1007/s13347-020-00415-6.
Ceccaroni, L., Woods, S.M., Sprinks, J., Wilson, S., Faustman, E.M., Bonn, A., Greshake Tzovaras, B., Subirats, L. and Kimura, A.H. (2021) 'Citizen science, health, and environmental justice', in Vohland, K., Land-Zandstra Ceccaroni, L., Lemmens, R., Perelló, J., Ponti, M., Samson, R. and Wagenknecht, K. (eds.) *The Science of Citizen Science*. Cham, Switzerland: Springer Nature, pp. 219–239. doi:10.1007/978-3-58278-4.
Cellan-Jones, R. (2014) 'Stephen Hawking warns artificial intelligence could end mankind', *BBC*, 2 December. Available at: https://www.bbc.com/news/technology-30290540.
Chakanya, N. (2018) 'The changing world of work: policy considerations for a better future for women in Africa', *BUWA!*, 19, pp. 44–48.
Chan, A., Okolo, C.T., Terner, Z. and Wang, A. (2021) 'The limits of global inclusion in AI development'. arXiv preprint arXiv:2102.01265.
Chander, S. (2021) 'Foreword', in Balayn, A. and Gürses, S., 'Beyond debiasing: regulating AI and its inequalities'. *EDRi Report*, pp. 8–9. Available at: https://edri.org/wp-content/uploads/2021/09/EDRi_Beyond-Debiasing-Report_Online.pdf.
Charalambous, S. (2021) 'Artificial intelligence and journalism: ready or not, here it comes', *Daily Maverick*, 28 September. Available at: https://www.dailymaverick.co.za/opinionista/2021-09-28-artificial-intelligence-and-journalism-ready-or-not-here-it-comes/.
Chen, I.Y., Joshi, S. and Ghassemi, M. (2020) 'Treating health disparities with artificial intelligence', *Nature Medicine*, 26(1), pp. 16–17. doi:10.1038/s41591-019-0649-2.
Chilembo, Z. and Tembo, Z. (2020) 'Opportunities and challenges of coordinating the implementation of E-government programmes in Zambia', *International Journal of Information Science*, 10(1), pp. 29–43. doi:10.5923/j.ijis.20201001.04.
Chimedza, T. (2019) 'Women, precarity, and the political economy of the Fourth Industrial Revolution', *BUWA!*, 19, pp. 96–101.
Chisale, S.S. (2018) 'Ubuntu as care: deconstructing the gendered Ubuntu', *Verbum et Ecclesia*, 39(1), pp. 1–8. doi:10.4102/ve.v39i1.1790.
Chiweshe, M.K. (2019) 'Policy briefing: women, power & policymaking. Fourth industrial revolution: what's in it for African women?'. *Centre for International Governance Innovation*. Available at: https://media.africaportal.org/documents/Chiweshe__Fourth_industrial_revolution.pdf.
Cissé, F. (2019) 'All is not lost: how Africa can ride the artificial intelligence wave', *The Africa Report*, 12 June. Available at: https://www.theafricareport.com/13917/how-africa-can-ride-the-artificial-intelligence-wave/.
Cisse, M. (2018) 'Look to Africa to advance artificial intelligence', *Nature*, 562(7728), pp. 461–462.
Clapham, C. (2020) 'Decolonising African studies?', *The Journal of Modern African Studies*, 58(1), pp. 137–153. doi:10.1017/S0022278X19000612.
Clement, G. (1996) *Care, Autonomy, and Justice*. Boulder, CO: Westview Press.
Coeckelbergh, M. (2021) 'AI for climate: freedom, justice, and other ethical and political challenges', *AI and Ethics*, 1(1), pp. 67–72. doi:10.1007/s43681-020-00007-2.
Cohen, P. (2014) 'Don't trouble yourself', *Boston Review*, 23 June. Available at: https://terpconnect.umd.edu/~pnc/BR2014.pdf.
Cohn, J. (2020) 'In a different code: artificial intelligence and the ethics of care', *The International Review of Information Ethics*, 28, pp. 1–7.

Cole, M., Cant, C., Spilda, F.U. and Graham, M. (2022) 'Politics by automatic means? A critique of artificial intelligence ethics at work', *Frontiers in Artificial Intelligence*, 5, Article 869114. doi:10.3389/frai.2022.869114.

Comaroff, J.L. and Comaroff, J. (2002) 'On personhood: an anthropological perspective from Africa', in Köpping, K.-P., Welker, M. and Wiehl, R. (eds.) *Die Outonome Person – Eine Europäische Erfindung?*. Munchën, Germany: Wilhelm Fink, pp. 67–82.

Cooper, M. (2020) 'Philosophical foundations of pluralism', *Pluralistic Practice*, 19 February. Available at: https://pluralisticpractice.com/2020/02/19/philosophical-foundations-of-pluralism/.

Constantinescu, M., Voinea, C., Uszkai, R. and Vică, C. (2021) 'Understanding responsibility in responsible AI: dianoetic virtues and the hard problem of context', *Ethics and Information Technology*. Available at: 10.1007/s10676-021-09616-9.

Copson, A. (2015) 'What is humanism?, in Copson, A. and Grayling, A.C. (eds.) *The Wiley Blackwell Handbook of Humanism*. Chichester, UK: John Wiley and Sons, pp. 1–34.

Cornell, D. and Van Marle, K. (2015) 'Ubuntu feminism: tentative reflections', *Verbum et Ecclesia*, 36(2), pp. 1–8. doi:10.4102/VE.V36I2.1444.

Correll, J., Park, B., Judd, C.M. and Wittenbrink, B. (2002) 'The police officer's dilemma: using ethnicity to disambiguate potentially threatening individuals', *Journal of Personality and Social Psychology*, 83(6), pp. 1314–1329. doi:10.1037/0022-3514.83.6.1314.

Crawford, K. (2021) *The Atlas of AI*. New Haven and London: Yale University Press. doi:10.12987/9780300252392.

Crosston, M. (2020) 'Cyber colonization: the dangerous fusion of artificial intelligence and authoritarian regimes', *Cyber, Intelligence, and Security Journal*, 4(1), pp. 149–171.

CSIR. (2022) 'Computer vision'. Available at: https://www.csir.co.za/computer-vision.

Cukurova, M., Luckin, R. and Kent, C. (2020) 'Impact of an artificial intelligence research frame on the perceived credibility of educational research evidence', *International Journal of Artificial Intelligence in Education*, 30, pp. 205–235. doi:10.1007/s40593-019-00188-w.

Daily Times reporter. (2017) '"Machine Learning May Erase Jobs", says Yudala', *Daily Times*, 28 August. Available at: https://dailytimes.ng/machine-intelligence-ai-may-erase-jobs-says-yudala/.

Darling, K. (2015) 'Who's Johnny? anthropomorphic framing in human-robot interaction, integration, and policy', in Lin, P., Bekey, G., Abney, K. and Jenkins, R. (eds.) *Robotic Ethics 2.0*. Oxford, UK: Oxford University Press, pp. 3–21.

Da Silveira, F., Lermen, F.H. and Amaral, F.G. (2021) 'An overview of agriculture 4.0 development: systematic review of descriptions, technologies, barriers, advantages, and disadvantages', *Computers and Electronics in Agriculture*, 189, p. 106405. doi:10.1016/j.compag.2021.106405.

De Jaegher, H. (2019) 'Loving and knowing: reflections for an engaged epistemology', *Phenomenology and the Cognitive Sciences*, 20(5), pp. 847–870. doi:10.1007/s11097-019-09634-5.

Delmolino, D. and Whitehouse, M. (2018) 'Responsible AI: a framework for building trust in your AI solutions', *Accenture Federal Services*.

Department of Higher Education and Training (2019) *Skills Supply and Demand in South Africa*. Pretoria, South Africa: Department of Higher Education and Training.

De Sousa Santos, B. (2018) *The End of the Cognitive Empire*. London, UK: Duke University Press. doi:10.1515/9781478002000.
De Vergès, M. (2021) 'How facial recognition technology is different in Africa', *Worldcrunch*, 13 January. Available at: https://worldcrunch.com/tech-science/how-facial-recognition-technology-is-different-in-africa.
Dias, G. (2020) 'South Africans are optimistic about AI, but unsure of how it affects them', *Paramount Insights*, 20 May. Available at: https://insights.paramount.com/post/south-africans-are-optimistic-about-ai-but-unsure-of-how-it-affects-them/.
Dicks, C. and Govender, P. (2019) *Feminist Visions of the Future of Work*. Berlin, Germany: Friedrich-Ebert-Stiftung.
Diga, K., Nwaiwu, F. and Plantinga, P. (2013) 'ICT policy and poverty reduction in Africa', *Info*, 15(4), pp. 114–127. doi:10.1108/info-05-2013-0032.
Dignum, V. (2021) 'The role and challenges of education for responsible AI', *London Review of Education*, 19(1), pp. 1–11. doi:10.14324/LRE.19.1.01.
Dolev, N. and Itzkovich, Y. (2020) 'In the AI era, soft skills are the new hard skills', in Amann, W. and Stachowicz-Stanusch, A. (eds.) *Artificial Intelligence and its Impact on Business*. Charlotte, NC: Information Age Publishing Inc., pp. 55–78.
Domingo, A. (2015) 'Migration as a global risk: the world economic forum and neoliberal discourse on demography', *Quetelet Journal*, 3(1), pp. 97–117. doi:10.14428/rqj2015.03.01.04.
DrivenData. (2021) 'Zamba'. Available at: https://github.com/drivendataorg/zamba.
Dube, S.I. (2021) 'The decoloniality of being political studies/science: legitimising a(nother) way of being', *Critical Studies in Teaching and Learning (CriSTaL)*, 9(1), pp. 58–77. doi:10.14426/cristal.v9i1.391.
Duberry, J. and Hamidi, S. (2021) 'Contrasted media frames of AI during the COVID-19 pandemic: a content analysis of US and European newspapers', *Online Information Review*, 45(4), pp. 758–776. doi:10.1108/OIR-09-2020-0393.
Dufva, T. and Dufva, M. (2019) 'Grasping the future of the digital society', *Futures*, 107, pp. 17–28. doi:10.1016/j.futures.2018.11.001.
Du Plessis, G.E. (2019) 'Gendered human (in) security in South Africa: what can Ubuntu feminism offer?', *Acta Academica*, 51(2), pp. 41–63. doi:10.18820/24150479/aa51i2.3.
Dursun, S. and Mankolli, H. (2021) 'The value of nature: virtue ethics perspective', *GNOSI: An Interdisciplinary Journal of Human Theory and Praxis*, 4(1), pp. 1–15.
Dusek, V. (2006) *Philosophy of Technology: An Introduction*. Malden, MA: Blackwell.
Du Toit, J.S. and Puttkammer, M.J. (2021) 'Developing core technologies for resource-scarce Nguni languages', *Information*, 12(12), p. 520. doi:10.3390/info12120520.
Dzwonkowska, D. (2018) 'Is environmental virtue ethics anthropocentric?', *Journal of Agricultural and Environmental Ethics*, 31(6), pp. 723–738. doi:10.1007/s10806-018-9751-6.
Eberhard, D.M., Gary, F.S. and Fennig, C.D. (eds.) (2019) *Ethnologue: Languages of the World* (22nd ed.). Dallas: SIL International.
Eckerson, W. (2007) 'Predictive analytics', *tdwi*, 10 May. Available at: http://tdwi.org/articles/2007/05/10/predictive-analytics.aspx.
Edeh, A. 2015. 'African humanism in achebe in relation to the west', *Open Journal of Philosophy*, 5(3), pp. 205–210. doi:10.4236/ojpp.2015.53025.
Effoduh, J.O. (2020) '7 ways that African states are legitimizing artificial intelligence', *Open Air*, 20 October. Available at: https://openair.africa/7-ways-that-african-states-are-legitimizing-artificial-intelligence/.

Egypt Today staff (2020) 'Egypt to introduce artificial intelligence in irrigation water management', *Egypt Today*, 10 August. Available at: https://www.egypttoday.com/Article/1/90613/Egypt-to-introduce-artificial-intelligence-in-irrigation-water-management.

Ehsan, U. and Riedl, M.O. (2020) 'Human-centered explainable AI: towards a reflective sociotechnical approach', in Stephanidis, C., Salvendy, G., Wei, J., Yamamoto, S., Mori, H., Meiselwitz, G., Nah, F. and Siau, K. (eds.) *International Conference on Human-Computer Interaction*. Cham, Switzerland: Springer, pp. 449–466. doi:10.1007/978-3-030-60117-1_33.

Ekanem, S.A. (2013) 'Technology and humanity: a humanist approach', *International Journal of Science and Research (IJSR)*, 2(1), pp. 77–82.

Ekdale, B. and Tully, M. (2019) 'African elections as a testing ground: comparing coverage of Cambridge analytica in Nigerian and Kenyan newspapers', *African Journalism Studies*, 40(4): pp. 27–43. doi:10.1080/23743670.2019.1679208.

Ekwe-Ekwe, H. (2012) 'What exactly does 'sub-Sahara Africa' mean?', *Pambazuka News*, 18 January. Available at: https://www.pambazuka.org/governance/what-exactly-does-%E2%80%98sub-sahara-africa%E2%80%99-mean.

Eleojo, E.F. (2014) 'Africans and African humanism: what prospects?', *American International Journal of Contemporary Research*, 4(1), pp. 297–308.

Elephant Listening Project. Cornell University. Available at: https://elephantlisteningproject.org/.

Elhag, M.M. and Abdelmawla, M.A. (2020) 'Gender-based assessment of science, technology and innovations ecosystem in Sudan', *African Journal of Rural Development*, 5(1), pp. 97–113.

Elisa, N. (2018) 'Could machine learning be used to address Africa's challenges?', *International Journal of Computer Applications*, 180(18), pp. 9–12.

Elwood, S. (2021) 'Digital geographies, feminist relationality, black and queer code studies: thriving otherwise', *Progress in Human Geography*, 45(2), pp. 209–228. doi:10.1177/0309132519899733.

Emejulu, A. (2014) 'Towards a radical digital citizenship'. Available at: https://digital.education.ed.ac.uk/showcase/towards-radical-digital-citizenship.

Emejulu, A. and McGregor, C. (2019) 'Towards a radical digital citizenship in digital education', *Critical Studies in Education*, 60(1), pp. 131–147. doi:10.1080/17508487.2016.1234494

Engster, D. and Hamington, M. (eds.) (2015) *Care Ethics and Political Theory*. Oxford, UK: Oxford University Press.

Esteva, A., Kuprel, B., Novoa, R.A., Ko, J., Swetter, S.M., Blau, H.M. and Thrun, S. (2017) 'Dermatologist-level classification of skin cancer with deep neural networks', *Nature*, 542(7639), pp. 115–118. doi:10.1038/nature21056.

Ethics Guidelines for Trustworthy AI. (2019) European Commission. Available at: https://ec.europa.eu/digital-single-market/en/news/ethicsguidelines-trustworthy-ai.

Evans, G. (2018) 'The unwelcome revival of "Race Science"', *The Guardian*, 2 March. Available at: https://www.theguardian.com/news/2018/mar/02/the-unwelcome-revival-of-race-science.

Ewuoso, C. and Fayemi, A.K. (2021) 'Transhumanism and African humanism: how to pursue the transhumanist vision without jeopardizing humanity', *Bioethics*, 35(7), pp. 634–645.

Ewuoso, C. and Hall, S. (2019) 'Core aspects of Ubuntu: a systematic review', *South African Journal of Bioethics and Law*, 12(2), pp. 93–103.

Ezeanya-Esiobu, C. (2019) *Indigenous Knowledge and Education in Africa*. Singapore: Springer Nature. doi:10.1007/978-981-13-6635-2.

Fairwork (2021) *Fairwork Cloudwork Ratings 2021: Labour Standards in the Platform Economy*. Oxford, United Kingdom: The Fairwork Project.

Famubode, V. (2018) Rising automation in sub-Saharan Africa: harnessing its opportunities through public policy', *SSRN*, 2 April. Available at: https://ssrn.com/abstract=3154359. doi:10.2139/ssrn.3154359.

Fanon, F. (1952) *Black Skin White Masks*. London, UK: Pluto Press.

Fast, E. and Horvitz, E. (2017) 'Long-term trends in the public perception of artificial intelligence', *Proceedings of the AAAI Conference on Artificial Intelligence (AAA-17)*, 31(1), pp. 963–969.

Fink, J. (2012) 'Anthropomorphism and human likeness in the design of robots and human-robot interaction', in Ge, S.S., Khatib, O., Cabibihan, J.J., Simmons, R. and Williams, M.A. (eds.) *Social Robotics*. Berlin, Heidelberg: Springer, pp. 199–208. doi:10.1007/978-3-642-34103-8_20.

Flanagan, F. (2017) 'Symposium on work in the "gig" economy: introduction', *The Economic and Labour Relations Review*, 28(3), pp. 378–381. doi:10.1177/1035304617724302.

Floridi, L. (2018) 'The ethics of artificial intelligence', in Franklin, D. (ed.) *Megatech: Technology in 2050*. London, UK: Profile Books, pp. 155–163.

Floridi, L. (2019) 'Translating principles into practices of digital ethics: five risks of being unethical', *Philosophy & Technology*, 32, pp. 185–193. doi:10.1007/s13347-019-00354-x.

Floridi, L. (2020) 'AI and its new winter: from myths to realities', *Philosophy & Technology*, pp. 33:1–3. doi:10.1007/s13347-020-00396-6.

Floridi, L. and Cowls, J. (2019) 'A unified framework of five principles for AI in society', *Harvard Data Science Review*, 11, pp. 1–15. doi: 10.1162/99608f92.8cd550d1.

Ford, M. (2013) 'Could artificial intelligence create an unemployment crisis?', *Communications of the ACM*, 56(7), pp. 1–3. doi:10.1145/2483852.2483865.

Forsyth, D.A. and Ponce, J. (2003) *Computer Vision: A Modern Approach* (2nd ed.). Upper Saddle River, NJ: Prentice Hall.

Frey, C.B. and Osborne, M.A. (2017) 'The future of employment: how susceptible are jobs to computerisation?', *Technological Forecasting and Social Change*, 114, pp.254–280. doi:10.1016/j.techfore.2016.08.019.

Friederici, N., Wahome, M. and Graham, M. (2020) *Digital Entrepreneurship in Africa: How a Continent is Escaping Silicon Valley's Long Shadow*. Cambridge, MA: The MIT Press.

Gabriel, I. (2020) 'Artificial intelligence, values, and alignment', *Minds and Machines*, 30(3), pp. 411–437. doi:10.1007/s11023-020-09539-2.

Gabriel, I. and Ghazavi, V. (2021) 'The challenge of value alignment: from fairer algorithms to AI safety'. arXiv preprint arXiv:2101.06060.

Gadzala, A. (2018) 'Coming to life: artificial intelligence in Africa', *Atlantic Council*, 14 November. Available at: https://www.atlanticcouncil.org/in-depth-research-reports/issue-brief/coming-to-life-artificial-intelligence-in-africa/.

Gaffley, M.M., Adams, R. and Shyllon, O. (2022) 'Artificial intelligence. African insight. A research summary of the ethical and human rights implications of AI in Africa. *HSRC & Meta AI and Ethics Human Rights Research Project for Africa – Synthesis Report*. Available at: https://africanaiethics.com/wp-content/uploads/2022/02/Artificial-Intelligence-African-Insight-Report.pdf.

Gagliardone, I., Kalemera, A., Kogen, L., Nalwoga, L., Stremlau, N. and Wakabi, W. (2015) *In Search of Local Knowledge on ICTs in Africa*, Working Paper, January 2015. University of Pennsylvania: CGCS. doi:10.5334/sta.fv.
Gallien, C. (2020) 'A decolonial turn in the humanities', *Alif: Journal of Comparative Poetics*, (40), pp. 28–58.
Galton, F. (1901) 'The possible improvement of the human breed under the existing conditions of law and sentiment', *Nature*, pp. 659–665.
Garbe, L. (2020) 'What we do (not) know about internet shutdowns in Africa', in Cheeseman, N. and Garbe, L. (eds.) *Decoding Digital Democracy in Africa*. A collaboration between Democracy in Africa (DIA) and the Digital Civil Society Lab (DCSL): Stanford PACS, pp. 31–35. Available at: https://pacscenter.stanford.edu/wp-content/uploads/2021/06/Decoding-Digital-Democracy-Booklet_Final_loRes-2.pdf.
Gaylard, R. (2004) 'Welcome to the world of our humanity: (African) humanism, Ubuntu and black South African writing', *Journal of Literary Studies*, 20(3–4), pp. 265–282.
Gbedomon, R.C. (2016) *Empowering Women in Technology: Lessons from Successful Woman Entrepreneur in Kenya Case Study No 10. A Joint Work by The African Capacity Building Foundation (ACBF) and The African Development Bank (AfDB)*. Available at: https://media.africaportal.org/documents/Case_study_10.pdf.
George, K.M. (2020) 'The African state in a wake of neoliberal globalization: a cog in a wheel or a wheel in a cog', *Journal of Research in Philosophy and History*, 3(2), pp. 32–51.
Gershgorn, D. (2019) 'Africa is building an ai industry that doesn't look like silicon valley', in *Medium*, 25 September. Available at: https://onezero.medium.com/africa-is-building-an-a-i-industry-that-doesnt-look-like-silicon-valley-72198eba706d.
Gill, K.S. (2017) 'Uncommon voices of AI', *AI & Society*, 32, pp. 475–482. doi:10.1007/s00146-017-0755-y.
Gilligan, C. (1982) *In a Different Voice: Psychological Theory and Women's Development*. Cambridge, MA: Harvard University Press.
Gillwald, A. (2020) 'Data, AI & society', *Research ICT Africa*, 10 March. Available at: https://researchictafrica.net/2020/03/10/data-ai-society/.
Givá, N. and Santos, L. (2020) 'A gender-based assessment of science, technology and innovation ecosystem in Mozambique', *African Journal of Rural Development*, 5(1), pp. 79–95.
Global Information Society Watch. (2019) *Artificial Intelligence: Human Rights, Social Justice and Development*. USA: APC.
Goodfellow, I., Bengio, Y. and Courville, A. (2016) *Deep Learning: Machine Learning Book*. Available at: http://www.deeplearningbook.org/.
Gopaldas, R. (2019) 'Digital dictatorship versus digital democracy in Africa', *Policy Insights* 75, October 2019. Johannesburg, South Africa: South African Institute of International Affairs. Available at: https://media.africaportal.org/documents/Policy-Insights-75-gopaldas.pdf.
Gottfredson, L.S. (1997) 'Mainstream science on intelligence: an editorial with 52 signatories, history, and bibliography', *Intelligence*, 24(1), pp. 13–23. doi: 10.1016/S0160-2896(97)90011-8.
Goyanes, R. (2018) 'Data for black lives is an organization using analytics as a tool for social change', *Garage Magazine*, 1 February. Available at: https://garage.vice.com/en_us/article/kzn4jn/data-for-black-lives-is-an-organization-using-analytics-as-a-tool-for-social-change.

Grau, C. (2011) 'There is no "I" in "Robot": robots and utilitarianism', in Anderson, M. and Anderson, S.L. (eds.) *Machine Ethics*. New York, NY: Cambridge University Press, pp. 451–463.

Gravett, W.H. (2020a) 'Digital coloniser? China and artificial intelligence in Africa', *SURVIVAL*, 62(6), pp. 153–178. doi:10.1080/00396338.2020.1851098.

Gravett, W.H. (2020b) 'Digital neo-colonialism: the Chinese model of internet sovereignty in Africa', *African Human Rights Law Journal*, 20(1), pp. 125–146. doi:10.17159/1996-2096/2020/v20n1a5.

Gray, M.L. and Suri, S. (2019) *Ghost Work: How to Stop Silicon Valley from Building a New Global Underclass*. Boston, MA: Houghton Mifflin.

Green, B. (2019) '"Good" isn't good enough', *Proceedings of the AI for Social.*

Good Workshop at NeurIPS 2019. Vancouver, Canada.

Greene, D., Hoffmann, A.L. and Stark, L. (2019) 'Better, nicer, clearer, fairer: a critical assessment of the movement for ethical artificial intelligence and machine learning', in Bui, T. (ed.) *Proceedings of the 52nd Hawaii International Conference on System Sciences*. Hawaii: Hawaii International Conference On System Sciences, pp. 2122–2131. doi:0.24251/HICSS.2019.258.

Guanah, J. and Ijeoma, O.B.I. (2020) 'Artificial intelligence and its reportage in select Nigerian newspapers: a content analysis', *International Journal of Language and Literary Studies*, 2(2), pp. 45–61. doi:10.36892/ijlls.v2i2.298.

Guerrini, C.J., Crossnohere, N.L., Rasmussen, L. and Bridges, J.F. (2021) 'A best-worst scaling experiment to prioritize concern about ethical issues in citizen science reveals heterogeneity on people level v. data-level issues', *Scientific Reports*, 11(1), pp. 1–9. doi:10.1038/s41598-021-96743-4.

Guo, J. and Li, B. (2018) 'The application of medical artificial intelligence technology in rural areas of developing countries', *Health Equity*, 2(1), pp. 174–181. doi:10.1089/heq.2018.0037.

Gurumurthy, A. and Chami, N. (2019) 'The wicked problem of AI governance', *Artificial Intelligence in India*, 2. Available at: http://library.fes.de/pdf-files/bueros/indien/15763.pdf.

Gwagwa, A., Kachidza, P., Siminyu, K. and Smith, M. (2021a) 'Responsible artificial intelligence in Sub-Saharan Africa: landscape and general state of play', *International Development Research Centre (IDRC)*, 29 March. AI4D.

Gwagwa, A., Kazim, E., Kachidza, P., Hilliard, A., Siminyu, K., Smith, M. and Shawe-Taylor, J. (2021b) 'Road map for research on responsible artificial intelligence for development (AI4D) in African countries: the case study of agriculture', *Patterns*, 2(12), p. 100381. doi:10.1016/j.patter.2021.100381.

Gwagwa, A., Kraemer-Mbula, E., Rizk, N., Rutenberg, I. and De Beer, J. (2020) 'Artificial intelligence (AI) deployments in Africa: benefits challenges and policy dimensions', *The African Journal of Information and Communication*, 26, pp.1–28. doi:10.23962/10539/30361.

Hadzi, A. and Roio, D. (2019) 'Restorative justice in artificial intelligence crimes', *Spheres: Journal for Digital Cultures*, (5), pp.1–18.

Hagendorff, T. (2020) 'The ethics of AI ethics: an evaluation of guidelines', *Minds and Machines*, 30(1), pp. 99–120. doi:10.1007/s11023-020-09517-8.

Hagerty, A. and Rubinov, I. (2019) 'Global AI ethics: a review of the social impacts and ethical implications of artificial intelligence'. arXiv preprint arXiv:1907.07892.

Haidt, J. (2003) 'The Moral Emotions', in Davidson, R.J., Scherer, K.R. and Goldsmith, H.H. (eds.) *Handbook of Affective Sciences*. Oxford, UK: Oxford University Press, pp. 852–870.

Hall, D., Du Toit, L. and Louw, D. (2013) 'Feminist ethics of care and Ubuntu', *Obstetrics and Gynaecology Forum*, 23(1), pp. 29–33.

Halwani, R. (2003) 'Care ethics and virtue ethics', *Hypatia*, 18(3), pp. 161–192. doi:10.1111/j.1527-2001.2003.tb00826.x.

Hameiri, S. and Jones, L. (2020) *Debunking the Myth of "Debt-trap Diplomacy:: How Recipient Countries Shape China's Belt and Road Initiative*, Research Paper, August 2020. London, UK: Chatham House. Available at: https://www.chathamhouse.org/sites/default/files/2020-08-25-debunking-myth-debt-trap-diplomacy-jones-hameiri.pdf.

Hamington, M. and Sander-Staudt, M. (eds.) (2011) *Applying Care Ethics to Business (Vol. 34)*. Berlin, Germany: Springer Science & Business Media.

Han, J., Kamber, M. and Pei, J. (2012) *Data Mining: Concepts and Techniques* (3rd ed). Burlington, NJ: Elsevier Science.

Hao, K. (2020) 'The problems AI has today go back centuries', *MIT Technology Review*, 31 July. Available at: https://www.technologyreview.com/2020/07/31/1005824/decolonial-ai-for-everyone/.

Hargittai, E. (2002) 'Second-level digital divide: differences in people's online skills', *First Monday*, 7(4), pp. 1–14. Available at: http://firstmonday.org/article/view/942/864. doi:10.5210/fm.v7i4.942.

Harrison, G.P. (2020) 'Science and race: what we really know', *Skeptic (Altadena, CA)*, 25(3), pp. 24–30.

Hart, M.A. (2010) 'Indigenous worldviews, knowledge, and research: the development of an indigenous research paradigm', *Journal of Indigenous Social Development*, 1(1), pp. 1–16.

Harvey, D. (2004) 'The "new" imperialism: accumulation by dispossession', *Socialist Register* 40, pp. 63–87.

Heinisch, B., Oswald, K., Weißpflug, M., Shuttleworth, S. and Belknap, G. (2021) 'Citizen humanities', in Vohland, K., Land-Zandstra Ceccaroni, L., Lemmens, R., Perelló, J., Ponti, M., Samson, R. and Wagenknecht, K. (eds.) *The Science of Citizen Science*. Cham, Switzerland: Springer Nature, pp. 97–118. doi:10.1007/978-3-58278-4.

Held, V. (ed.) (1995) *Justice and Care: Essential Readings in Feminist Ethics*. Boulder, CO: Westview Press.

Held, V. (2006) *The Ethics of Care: Personal, Political, and Global*. New York, NY: Oxford University Press.

Heldring, L. and Robinson, J. (2013) *Colonialism and Development in Africa. African Economic History*, Working Paper Series, 5(2012) Available at: http://citeseerx.ist.psu.edu/viewdoc/download?doi=10.1.1.673.1830&rep=rep1&type=pdf.

Henry, N., Vasil, S. and Witt, A. (2021) 'Digital citizenship in a global society: a feminist approach', *Feminist Media Studies*, pp. 1–18. doi:10.1080/14680777.2021.1937269.

Herring, J. (2014) 'The disability critique of care', *Elder Law Review*, 8, pp. 1–15.

Herring, J. (2017) 'Compassion, ethics of care and legal rights', *International Journal of Law in Context*, 13(2), pp. 158–171. doi:10.1017/S174455231700009X.

Hlalele, D.J. (2019) 'Indigenous knowledge systems and sustainable learning in rural South Africa', *Australian and International Journal of Rural Education*, 29(1), pp. 88–100.

HLEGAI [High Level Expert Group on Artificial Intelligence] (2019) 'Ethics guidelines for trustworthy AI', *European Commission*, 8 April. Available at: https://www.aepd.es/sites/default/files/2019-12/ai-ethics-guidelines.pdf.

Hogarth, I. (2018), 'AI nationalism', *IanHogarth.com*, 13 June. Available at: https://www.ianhogarth.com/blog/2018/6/13/ai-nationalism.

HolonIQ Report (2020) *The 2020 AI Strategy Landscape: 50 National Artificial Intelligence Strategies Shaping the Future of Humanity*. Available at: https://www.holoniq.com/notes/50-national-ai-strategies-the-2020-ai-strategy-landscape/.

Holzmeyer, C. (2021) 'Beyond "AI for social good" (AI4SG): social transformations – not tech-fixes – for health equity', *Interdisciplinary Science Reviews*, 46(1–2), pp. 94–125. doi:10.1080/03080188.2020.1840221.

Horsthemke, K., (2008) 'The idea of indigenous knowledge', *Archaeologies*, 4(1), pp. 129–143. doi:10.1007/s11759-008-9058-8.

Hruby, A. (2018) 'Clash of the US and Chinese tech giants in Africa', *Financial Times*, 4 September, Available at: https://www.ft.com/content/ff7941a0-b02d-11e8-99ca-68cf89602132.

Hurtado, M. (2016) 'The ethics of super intelligence', *International Journal of Swarm Intelligence and Evolutionary Computing*, 5(137), pp. 1–8. doi:10.4172/2090-4908.1000137.

IDRC. (2021) 'Meet the AI4D Africa partners leading policy research', *IDRC*, 25 November. Available at: https://www.idrc.ca/en/research-in-action/meet-ai4d-africa-partners-leading-policy-research.

Igwe, O. (2002) *Politics and Globe Dictionary*. Enugu: Jamoe Enterprises.

Ihejirika, C.I. (2015) 'Ethico-epistemological implications of artificial intelligence for humanity', *Sophia: An African Journal of Philosophy*, 16(1), pp. 190–201.

International Humanist and Ethical Union (IHEU). (1996) *IHEU Minimum Statement on Humanism*. Humanists International, General Assembly. Available at: https://humanists.international/policy/iheu-minimum-statement-on-humanism/.

Jackson, T. (2019) 'Senegal becomes 2nd African nation to pass startup act', *Disrupt Africa*, 30 December. Available at: https://disrupt-africa.com/2019/12/30/senegal-becomes-2nd-african-nation-to-pass-startup-act/.

Jaiyeola, E.O. (2020) 'Patriarchy and colonization: the "brooder house" for gender inequality in Nigeria', *Journal of Research on Women and Gender*, 10, pp. 3–22.

Janiesch, C., Zschech, P. and Heinrich, K. (2021) 'Machine learning and deep learning', *Electron Markets*, 31, 685–695. doi:10.1007/s12525-021-00475-2.

Jansen, J. (2017) *As By Fire: The End of the South African University*. Cape Town, South Africa: Tafelberg.

Jansen Van Vuuren, A.M.J. and Celik, T. (2019) 'Fake narratives, dominant discourses: the role and influence of algorithms on the online South African land reform debate', in Van der Waag-Cowling, N. and Leenen, L. (eds.) *ICCWS 2019 14th International Conference on Cyber Warfare and Security: ICCWS 2019*. Reading, UK: Academic Conferences and Publishing Limited, pp. 142–147.

Jantijies, M. (2019) 'The role of educational institutions in closing STEM education gaps', in Sey, A. and Hafkin, N. (eds.) *Taking Stock: Data and Evidence on Gender Equality in Digital Access, Skills and Leadership*. Tokyo, Japan: United Nations University, pp. 274–280.

Jili, B. (2020) 'Surveillance technology a concern for many in Africa', *New Africa Daily*, 29 December. Available at: https://newafricadaily.com/surveillance-technology-concern-many-africa.

Jobin, A., Ienca, M. and Vayena, E. (2019) 'The global landscape of AI ethics guidelines', *Nature Machine Intelligence*, 1(9), pp. 389–399.

Johnson, A.T. and Mbah, M.F. (2021) '(Un)subjugating indigenous knowledge for sustainable development: considerations for community-based research in African Higher education', *Journal of Comparative and International Higher Education*, 13, p. 3707. doi:10.32674/jcihe.v13iSummer.3707.

Jones, L. and Hameiri, S. (2020) *Debunking the Myth of "Debt-trap Diplomacy", Research Paper*, August 2020. London, UK: Chatham House. Available at: https://www.chathamhouse.org/sites/default/files/2020-08-19-debunking-myth-debt-trap-diplomacy-jones-hameiri.pdfgolp.

Jones, R.H. and Hafner, C.A. (2021) *Understanding Digital Literacies: A Practical Introduction*. London, UK: Routledge. doi:10.4324/9781003177647.

Joyce, K., Smith-Doerr, L., Alegria, S., Bell, S., Cruz, T., Hoffman, S.G., Noble, S.U. and Shestakofsky, B. (2021) 'Toward a sociology of artificial intelligence: a call for research on inequalities and structural change', *Socius*, 7, pp. 1–11. doi:10.1177/2378023121999581.

Kamau, C.G. and Ilamoya, S. (2021) 'Accounting profession: steps into the future (African perspective)', *SSRN*, 21 October. Available at: https://ssrn.com/abstract=3946885. doi:10.2139/ssrn.3946885.

Kamulegeya, L.H., Ssebwana, J., Abigaba, W., Bwanika, J.M. and Musinguzi, D. (2019) 'Mobile health in uganda: a case study of the medical concierge group', *East Africa Science*, 1(1), pp. 9–14.

Kaplan, A. and Haenlein, M. (2020) 'Rulers of the world, unite! The challengers and opportunities of artificial intelligence', *Business Horizons*, 63, pp. 37–50. doi:10.1016.j.bushor.2019.09.003.

Kapuire, G.K., Cabrero, D.G., Stanley, C. and Winschiers-Theophilus, H. (2015) 'Framing technology design in Ubuntu: two locales in pastoral Namibia', in Ploderer, B., Carter, M., Gibbs, M.R., Smith, W. and Vetere, F. (eds.) *Proceedings of the Annual Meeting of the Australian Special Interest Group for Computer Human Interaction*. Australia: ACM, pp. 212–216. doi:10.1145/2838739.2838788.

Karekwaivanane, G. and Msonza, N. (2021) 'Zimbabwe digital rights landscape report', in Roberts, T. (ed.) *Digital Rights in Closing Civic Space: Lessons from Ten African Countries* Brighton, UK: Institute of Development Studies, pp. 43–60. doi:10.19088/IDS.2021.003.

Karimi, H. and Pina, A. (2021) 'Strategically addressing the soft skills gap among STEM undergraduates', *Journal of Research in STEM Education*, 7(1), pp. 21–46. doi:10.51355/jstem.2021.99.

Karlsson, M. and Edvardsson Björnberg, K. (2021) 'Ethics and biodiversity offsetting', *Conservation Biology*, 35(2), pp. 578–586. doi:10.1111/cobi.13603.

Kasperson, R.E., Renn, O., Slovic, P., Brown, H.S., Emel, J., Goble, R., Kasperson, J.X. and Ratick, S. (1988) 'The social amplification of risk: a conceptual framework', *Risk Analysis*, 8(2), pp. 177–187. doi:10.1111/j.1539-6924.1988.tb01168.x.

Kassongo, R.F., Tucker, W.D. and Pather, S. (2018) 'Government facilitated access to ICTs: adoption, use and impact on the well-being of indigent South Africans', *2018 IST-Africa Week Conference (IST-Africa)*. IEEE, Page-1.

Kaushal, A., Altman, R. and Langlotz, C. (2020) 'Geographic distribution of US cohorts used to train deep learning algorithms', *Jama*, 324(12), pp. 1212–1213. doi:10.1001/jama.2020.12067.

Kazim, E. and Koshiyama, A.S. (2021) 'A high-level overview of AI ethics', *Patterns*, 2(9), pp. 1–12. doi:10.1016/j.patter.2021.100314.

Keevy, I. (2009) 'Ubuntu versus the core values of the South African constitution', *Journal for Juridical Science*, 34(2), pp. 19–58.

Kehdinga, G.F. (2019) 'Education and the Fourth Industrial Revolution: challenges and possibilities for engineering education', *International Journal of Mechanical Engineering and Technology*, 10(8), pp. 271–284.

Kehdinga, G.F. (2020) 'Theorising the itinerant curriculum as the pathway to relevance in african higher education in the era of the Fourth Industrial Revolution', *International Journal of Education and Practice*, 8(2), pp. 248–256. doi:10.18488/journal.61.2020.82.248.256.

Kemeh, E. (2018) 'Ubuntu as a framework for the adoption and use of E-learning in Ghanaian public universities. In Takyi-Amoako, E.J. and Assié-Lumumba, N.D.T. (eds.) *Re-Visioning Education in Africa: Ubuntu-Inspired Education for Humanity*. Cham, Switzerland: Palgrave Macmillan, pp. 75–192. doi:10.1007/978-3-319-70043-4.

Kemp, S. (2020) *Digital 2020: South Africa*. Available at: https://datareportal.com/reports/digital-2020-south-africa.

Khakurel, J., Penzenstadler, B., Porras, J., Knutas, A. and Zhang, W. (2018) 'The rise of artificial intelligence under the lens of sustainability', *Technologies*, 6(4), pp. 1–18. https://doi.org/10.3390/technologies6040100.

Kiemde, S.M.A. and Kora, A.D. (2021) 'Towards an ethics of AI in Africa: rule of education', *AI and Ethics*, pp.1–6. doi:10.1007/s43681-021-00106-8.

Kisusu, R.W., Tongori, S.T. and Madiany, D.O. (2020) 'Implications of the Fourth Industrial Revolution on gender participation: a case in Tanzania, sub-Saharan Africa', *International Journal of Political Activism and Engagement (IJPAE)*, 7(4), pp. 13–25. https://doi.org/10.4018/IJPAE.2020100102.

Knight, W. (2019) 'African AI experts get excluded from a conference – again', *Wired*, 8 November. Available at: https://www.wired.com/story/african-ai-experts-get-excluded-from-a-conference-again/.

Knopf, K. (2015) 'The turn toward the indigenous: knowledge systems and practices in the academy', *Amerikastudien/American Studies*, pp. 179–200.

Kolawole, M.M. (2004) 'Re-conceptualizing African gender theory: feminism, womanism, and the arere metaphor', in Signe, A. (ed.) *Re-thinking Sexualities in Africa*. Uppsala: The Nordic Africa Institute, pp. 251–268.

Kosse, F. and Tincani, M.M. (2020) 'Prosociality predicts labor market success around the world', *Nature Communications*, 11(1), p. 5298. doi:10.1038/s41467-020-19007-1.

Kothari, A. and Cruikshank, S.A. (2021) 'Artificial intelligence and journalism: an agenda for journalism research in Africa', *African Journalism Studies*, pp. 17–33. doi:10.1080/23743670.2021.1999840.

KPMG. (2021) 'Algorithmic bias and financial services'. Available at: https://www.finastra.com/sites/default/files/documents/2021/03/market-insight_algorithmic bias-financialservices.pdf.

Kranzberg, M. (1986) 'Technology and history: "Kranzberg's Laws"', *Technology and Culture*, 27 (3): 544. doi:10.2307/3105385.

Kroes, P., Franssen, M., Poel, I.V.D. and Ottens, M. (2006) 'Treating socio-technical systems as engineering systems: some conceptual problems', *Systems Research and Behavioral Science: The Official Journal of the International Federation for Systems Research*, 23(6), pp. 803–814. doi:10.1002/sres.703.

Krönke, M. (2020) *Africa's Digital Divide and the Promise of E-learning*, Afrobarometer Policy Paper No. 66, June 2020. Available at: https://media.africaportal.org/documents/pp66-africas_digital_divide_and_the_promise_of_e-learning-afrobarometer_policy_s1oxzDa.pdf.

Krzywinski, M. and Cairo, A. (2013) 'Points of view: storytelling', *Nature Methods*, 10(8), p. 687. doi:10.1038/nmeth.2571.

Lakmeeharan, K., Manji, Q., Nyairo, R. and Poeltner, H. (2020) *Solving Africa's Infrastructure Paradox*. McKinsey & Company, 6. Available at: https://www.mckinsey.com/business-functions/operations/our-insights/solving-africas-infrastructure-paradox.

Lamola, M.J. (2021) 'The future of artificial intelligence, posthumanism and the inflection of Pixley Isaka Seme's African humanism', *AI & Society*, pp. 1–11. doi:10.1007/s00146-021-01191-3.

Landry, D. and Portelance, G. (2021) *More Problems More Money? Does China Lend More to African Countries with Higher Credit Risk Levels?*, Working Paper 568, March 2021. Washington, DC: Center for Global Development. Available at: https://www.cgdev.org/sites/default/files/more-problems-more-money-does-china-lend-more-african-countries-higher-credit-risk.pdf.

Langat, S.K., Mwakio, P.M. and Ayuku, D. (2020) 'How Africa should engage Ubuntu ethics and artificial intelligence', *Journal of Public Health International*, 2(4), pp. 20–25. doi:10.14302/issn.2641-4538.jphi-20-3427.

Larson, E.J. (2021) *The Myth of Artificial Intelligence: Why Computers Can't Think the Way We Do*. Cambridge, MA: Belknap Press.

Lee, A. (2018a) *Humanism and Empire: The Imperial Ideal in Fourteenth-Century Italy*. UK: Oxford University Press.

Lee, M.K., Kusbit, D., Metsky, E. and Dabbish, L. (2015) 'Working with Machines: The Impact of Algorithmic and Data-Driven Management on Human Workers', in *Proceedings of the 33rd Annual ACM Conference on Human Factors in Computing Systems*. Available at: https://www.cs.cmu.edu/~mklee/materials/Publication/2015-CHI_algorithmic_management.pdf.

Lee, I. and Shin, Y.J. (2020) 'Machine learning for enterprises: applications, algorithm selection, and challenges', *Business Horizons, Elsevier*, 63(2), pp. 157–170. doi: 10.1016/j.bushor.2019.10.005.

Lee, N.T. (2018b) 'Detecting racial bias in algorithms and machine learning', *Journal of Information, Communication and Ethics in Society*, 16(3), pp. 252–260. doi:10.1108/JICES-06-2018-0056.

Leonardi, P.M. (2012) 'Materiality, sociomateriality, and socio-technical systems: what do these terms mean? How are they different? do we need them?', in Leonardi, P.M., Nardi, B.A. and Kallinikos, J. (eds.) *Materiality and Organizing: Social Interaction in a Technological World*. Oxford, UK: Oxford University Press, pp. 25–48.

Letsebe, K. (2018) 'Women remain underrepresented in the ICT Field', *ITWeb*, 8 March. Available at: https://www.itweb.co.za/content/G98YdMLxaPAMX2PD.

Levine, P. (2017) *Eugenics: A Very Short Introduction* (Vol. 495). Oxford, UK: Oxford University Press.

Libe, P. (2020) 'Opening the Tech sector to Africa's women', *CGTN*, 26 July. Available at: https://news.cgtn.com/news/2020-07-26/Opening-the-tech-sector-to-Africa-s-women-Sq0u0p0sTK/index.html.

Links, F. (2021) 'AI, biometrics and no protection from abuse', *Namibian*, 24 February. Available at: https://www.namibian.com.na/209054/archive-read/AI-Biometrics-and-No-Protection-from-Abuse.

Lin, S.Y., Mahoney, M.R. and Sinsky, C.A. (2019) 'Ten ways artificial intelligence will transform primary care', *Journal of General Internal Medicine*, 34(8), pp. 1626–1630. doi:10.1007/s11606-019-05035-1.

Lindén, C.-G. (2017) 'Algorithms for journalism: the future of news work', *Journal of Medical Insight*, 4(1), pp. 60–76. doi:10.5617/jmi.v4i1.2420.

Lisinge, R.T. (2020) 'The belt and road initiative and Africa's regional infrastructure development: implications and lessons', *Transnational Corporations Review*, 12(4), pp. 425–438. doi:10.1080/19186444.2020.1795527.

Liu, X., Nourbakhsh, A., Li, Q., Shah, S., Martin, R. and Duprey, J. (2017) 'Reuters tracer: toward automated news production using large scale social media data', in *2017 IEEE International Conference on Big Data (Big Data). 2017 IEEE International Conference on Big Data (Big Data)*, IEEE, pp. 1483–1493. doi:10.1109/BigData.2017.8258082.

Lo, C. (2021) 'The great instability of digital disruption (i): China's digital belt & road initiative', *BNP Paribas*, 18 February. Available at: https://china.bnpparibas-am.com/2021/02/18/the-great-instability-of-digital-disruption-i-chinas-digital-belt-road-initiative/.

Lokanathan, V. (2020) 'China's belt and road initiative: implications in Africa', *Observer Research Foundation Issue*, 395, pp. 1–12. Available at: https://www.orfonline.org/wp-content/uploads/2020/08/ORF_IssueBrief_395_BRI-Africa.pdf.

Louw, D. (2010) *Power Sharing and the Challenge of Ubuntu Ethics*. Stellenbosch, South Africa: University of Stellenbosch.

Luengo-Oroz, M. (2019) 'Solidarity should be a core ethical principle of AI', *Nature Machine Intelligence*, 1(11), pp. 494–494. doi:10.1038/s42256-019-0115-3.

Lund, B.D., Omame, I., Tijani, S. and Agbaji, D. (2020) 'Perceptions toward artificial intelligence among academic library employees and alignment with the diffusion of innovations' adopter categories', *College & Research Libraries*, 81(5), pp. 865–882.

Lupton, D. (2017) 'Digital bodies', in Andrews, D., Silk, M. and Thorpe, H. (eds.) *Handbook of Physical Cultural Studies*. London, UK: Routledge, pp. 200–208. doi:10.4324/9781315745664.

Lupton, D. (2021) '"Flawed", "Cruel" and "Irresponsible": the framing of automated decision-making technologies in the Australian Press', *SSRN*, 18 April. Available at: https://ssrn.com/abstract=3828952. doi:10.2139/ssrn.3828952.

Lutz, C. (2019) 'Digital inequalities in the age of artificial intelligence and big data', *Human Behavior and Emerging Technologies*, 1(2), pp. 141–148. doi:10.1002/hbe2.140.

Ly, R. (2021) 'Machine learning challenges and opportunities in the African agricultural sector – a general perspective'. arXiv preprint arXiv:2107.05101.

Lynn, R. and Vanhanen, T. (2002) *IQ and the Wealth of Nations*. Westport, Connecticut and London: Praeger.

Lyons, A., Kass-Hanna, J., Zucchetti, A. and Cobo, C. (2019) 'Leaving no one behind: measuring the multidimensionality of digital literacy in the age of AI and other transformative technologies', *T20 Policy Brief*, pp. 1–17. Available at: https://t20japan.org/policy-brief-multidimensionality-digital-literacy.

Mabvurira, V. (2020) 'Hunhu/Ubuntu philosophy as a guide for ethical decision making in social work', *African Journal of Social Work*, 10(1), pp. 73–77.

Magubane, T. (2021) 'Technology and law: the use of artificial intelligence and 5G to access the courts in Africa', *Young African Leaders Journal of Development*, 3(1), Article 11. Available at: https://digitalcommons.kennesaw.edu/yaljod/vol3/iss1/11.

Mahomed, S. (2018) 'Healthcare, artificial intelligence and the Fourth Industrial Revolution: ethical, social and legal considerations', *South African Journal of Bioethics and Law*, 11(2), pp. 93–95. doi:10.7196/SAJBL.2018.v11i2.664.

Maiyane, K. (2019) 'Ethics of artificial intelligence: virtue ethics as a solution to artificial moral reasoning in the context of lethal autonomous weapon systems'. Available at: http://ceur-ws.org/Vol-2540/FAIR2019_paper_22.pdf.

Majama, K. (2019) 'African women face widening technology gap', *AfriSIG*, 1 April. Available at: https://afrisig.org/2019/04/01/african-women-face-widening-technology-gap/.

Malanga, D.F. (2019) 'Framing the impact of artificial intelligence on protection of women's rights in Malawi', in Finlay, A. (ed.) *Global Information Watch 2019. Artificial Intelligence: Human Rights, Social Justice and Development*. USA: APC.

Mannion, C. (2020) 'Data imperialism: the GDPR's disastrous impact on Africa's E-commerce markets', *Vanderbilt Journal of Transnational Law*, 53(2), pp. 685–712.

Manyonganise, M. (2015) 'Oppressive and liberative: a Zimbabwean woman's reflections on Ubuntu', *Verbum et Ecclesia* 36(2), pp. 1438–1445. doi:10.4102/ve.v36i2.1438.

Marivate, V. (2020) *Why African Natural Language Processing Now? A View from South Africa# AfricaNLP*, MISTRA Working Paper, 2 November. Available at: https://mistra.org.za/wp-content/uploads/2020/11/4IR-Working-Paper_Dr-Vukosi-Marivate_20201102.pdf.

Markauskaite, L., Marrone, R., Poquet, O., Knight, S., Martinez-Maldonado, R., Howard, S., Tondeur, J., De Laat, M., Shum, S.B., Gašević, D. and Siemens, G. (2022) 'Rethinking the entwinement between artificial intelligence and human learning: what capabilities do learners need for a world with AI?', *Computers and Education: Artificial Intelligence*, 3, p.100056. doi:10.1016/j.caeai.2022.100056.

Markovic, M. (2019) 'Rise of the robot lawyers?', *Arizona Law Review*, 61(325), pp. 325–350.

Marr, B. (2018) 'The key definitions of artificial intelligence that explain its importance', *Forbes*, 14 February. Available at: https://www.forbes.com/sites/bernardmarr/2018/02/14/the-key-definitions-of-artificial-intelligence-ai-that-explain-its-importance/?sh=4026037a4f5d.

Martinho, A., Herber, N., Kroesen, M. and Chorus, C. (2021) 'Ethical issues in focus by the autonomous vehicles industry', *Transport Reviews*, pp. 1–22. doi:10.1080/01441647.2020.1862355.

Marx, K. (1852) 'The eighteenth Brumaire of Louis Bonaparte', in *The Collected Works of Karl Marx and Friedrich Engels (Volume 11)*. Charlottesville, VA: InteLex Corporation, pp. 103–197.

Masinde, M. (2015) 'An innovative drought early warning system for sub-Saharan Africa: integrating modern and indigenous approaches', *African Journal of Science, Technology, Innovation and* Development, 7(1), pp. 8–25. doi:10.1080/20421338.2014.971558.

Maumela, T., Ṋelwamondo, F. and Marwala, T. (2020) 'Introducing Ulimisana optimization algorithm based on Ubuntu philosophy', *IEEE Access*, 8, pp. 179244–179258. doi:10.1109/ACCESS.2020.3026821.

Mazzocchi, F. (2006) 'Western science and traditional knowledge: despite their variations, different forms of knowledge can learn from each other', *EMBO Reports*, 7(5), pp. 463–466. doi:10.1038/sj.embor.7400693.

Mbembe, A. (2015) 'Decolonizing knowledge and the question of the archive'. Available at: https://wiser.wits.ac.za/sites/default/files/private/Achille%20Mbembe%20-%20Decolonizing%20Knowledge%20and%20the%20Question%20of%20the%20Archive.pdf.

McBride, N. and Liyala, S. (2021) 'Memoirs from Bukhalalire: a poetic inquiry into the lived experience of M-PESA mobile money usage in rural Kenya', *European Journal of Information Systems*, pp. 1–22. doi:10.1080/0960085X.2021.1924088.

McClure, P.K. (2018) '"You're fired", says the robot: the rise of automation in the workplace, technophobes, and fears of unemployment', *Social Science Computer Review*, 36(2), pp. 139–156. doi:10.1177/0894439317698637.

McDonald, H. (2019) 'The internet as an extension of colonialism', *The Security Distillery*, 4 December 2019. Available at: https://thesecuritydistillery.org/all-articles/the-internet-as-an-extension-of-colonialism.

McLennan, S., Lee, M.M., Fiske, A. and Celi, L.A. (2020) 'AI ethics is not a panacea', *The American Journal of Bioethics*, 20(11), pp. 20–22. doi:10.1080/15265161.2020.1819470.

Metz, T. (2007) 'Toward an African moral theory', *Journal of Political Philosophy*, 15(3), pp. 321–341.

Metz, T. (2011) 'Ubuntu as a moral theory and human rights in South Africa', *African Human Rights Law Journal*, 11(2), pp. 532–559.

Metz, T. (2012) 'Ethics in Africa and in Aristotle: some points of contrast: special theme articles', *Phronimon*, 13(2), pp. 99–117.

Metz, T. (2021) 'African reasons why artificial intelligence should not maximize utility', in Okyere-Manu, B.D. (ed.) *African Values, Ethics, and Technology: Questions, Issues, and Approaches*. Cham, Switzerland: Springer Nature, pp. 55–72. doi:10.1007/978-3-030-705503.

Metzinger, T. (2019) 'EU Guidelines: ethics washing made in Europe', *Der Tagesspiegel*, 8 April. Available at: https://www.tagesspiegel.de/politik/eu-guidelines-ethics-washing-made-in-europe/24195496.html.

Mhlambi, S. (2020) *From Rationality to Relationality: Ubuntu as an Ethical and Human Rights Framework for Artificial Intelligence Governance*, Carr Center for Human Rights Policy Discussion Paper Series, 9.

Mhlanga, D. (2020) 'Industry 4.0 in finance: the impact of artificial intelligence (AI) on digital financial inclusion', *International Journal of Financial Studies*, 8(45), pp. 1–14. doi:10.3390/ijfs8030045.

Mikolov, T., Chen, K., Corrado, G. and Dean, J. (2013) 'Efficient estimation of word representations in vector space', in *1st International Conference on Learning Representations, ICLR 2013 – Workshop Track Proceedings*. Available at: http://arxiv.org/abs/1301.3781.

Millington, A.K. (2017) 'How changes in technology and automation will affect the labour market in Africa', *K4D Helpdesk Report*. Brighton, UK: Institute of Development Studies. Available at: https://opendocs.ids.ac.uk/opendocs/handle/20.500.12413/13054.

Mittelstadt, B. (2019) 'Principles alone cannot guarantee ethical AI', *Nature Machine Intelligence*, 1(11), pp. 501–507. doi:10.1038/s42256-019-0114-4.

Mnyaka, M. and Motlhabi, M. (2009) 'Ubuntu and its socio-moral significance', in Murove, M.F. (ed.) *African Ethics: An Anthology of Comparative and Applied Ethics*. Scottsville, South Africa: University KwaZulu-Natal Press, pp. 63–84.

Mohamed, S. (2018). 'Decolonising Artificial Intelligence', *The Spectator*, 11 October. Available at: https://blog.shakirm.com/2018/10/decolonising-artificial-intelligence/.

Mohamed, S., Png, M.T. and Isaac, W. (2020) 'Decolonial AI: decolonial theory as sociotechnical foresight in artificial intelligence', *Philosophy & Technology*, 33(4), pp. 659–684. doi:10.1007/s13347-020-00405-8.

Mohri, M., Rostamizadeh, A. and Talwalkar, A. (2012) *Foundations of Machine Learning*. Cambridge, MA: MIT Press.

Moitse, M., Seabi, M., Mthethwa, N. and Charlie, C. (2018) *Developing an Economically Active Citizen During the Time of the Fourth Industrial Revolution: Research Report*. Johannesburg, South Africa: Kagiso Trust. Available at: https://www.kagiso.co.za/wp-content/uploads/2019/09/Development-of-economic-active-citizen_Research-report-_SPM-6-Sep-2018.pdf.

Molefe, M. (2021) 'Personhood, dignity, duties and needs in African philosophy', in Molefe, M. and Allsobrook, C. (eds.) *Towards an African Political Philosophy of Needs*. Cham, Switzerland: Palgrave Macmillan, pp. 57–86. doi:10.1007/978-3-030-64496-3_4.

Montreal Declaration for the Responsible Development of Artificial Intelligence. (2018) University of Montreal, Canada. Available at: https://docs.wixstatic.com/ugd/ebc3a3_c5c1c196fc164756afb92466c081d7ae.pdf.

Moore, J. (2019) 'AI for not bad', *Frontiers in Big Data*, 2, Article 32. doi:10.3389/fdata.2019.00032.

Moore, J.W. (2017) 'Anthropocenes & the capitalocene alternative', *Azimuth*, 9, pp. 71–79.

Moyo, A. (2019) 'South African accountants allay AI fears', *ITWeb Africa*. Available at: https://www.itweb.co.za/content/O2rQGqApeX6Md1ea (Accessed: 18 January 2019).

Mudau, J. and Mukonza, R.M. (2021) 'Scant penetration of women in the Fourth Industrial Revolution: an old problem in a new context', *African Journal of Gender, Society and Development*, 10(1), pp. 85–98.

Mudongo, O. (2020) 'Botswana's quest for Fourth Industrial Revolution, a delusion of grandeur?', *Research ICT Africa*, 21 January. Available at: https://researchictafrica.net/2020/01/21/botswanas-quest-for-fourth-industrial-revolution-4ir-a-delusion-of-grandeur/ (Accessed: 21 January 2020).

Mudongo, O. (2021) 'Africa's expansion of AI surveillance: regional gaps and key trends', *Research ITC Africa*, 26 February. Available at: https://media.africaportal.org/documents/AI-Surveillance_Policy-Brief_Oarabile_Final.pdf.

Mueller, M., Tippins, D. and Bryan, L. (2012) 'The future of citizen science', *Democracy & Education*, 20(1), pp. 1–12.

Mukhwana, A.M., Abuya, T., Matanda, D., Omumbo, J. and Mabuka, J. (2020) 'Factors which contribute to or inhibit women in science, technology, engineering, and mathematics in Africa', *Nairobi: African Academy of Sciences*. Available at: https://www.aasciences.africa/publications/factors-which-contribute-or-inhibit-women-science-technology-engineering-and.

Mulrean, C. (2020) 'Women in the Fourth Industrial Revolution: a gendered perspective on digitalization in Kenya, Nigeria and South Africa'. Unpublished MA dissertation. Centre International de Formation Européenne.

Munoriyarwa, A., Chiumbu, S. and Motsaathebe, G. (2021) 'Artificial intelligence practices in everyday news production: the case of South Africa's mainstream newsrooms', *Journalism Practice*, pp. 1–19. doi:10.1080/17512786.2021.1984976.

Mupangwa, W., Chipindu, L., Nyagumbo, I., Mkuhlani, S. and Sisito, G. (2020) 'Evaluating machine learning algorithms for predicting maize yield under conservation agriculture in Eastern and Southern Africa', *SN Applied Sciences*, 2(952), pp. 1–13. doi:10.1007/s42452-020-2711-6.

Mutch, A. (2013) 'Sociomateriality – taking the wrong turn?', *Information and Organization*, 23(1), pp. 28–40. doi:10.1016/j.infoandorg.2013.02.001.

Mutsvairo, B., Ragnedda, M. and Orgeret, K.S. (2021) 'Era or error of transformation? Assessing afrocentric attributes to digitalization', *Information, Communication & Society*, 24(3), pp. 295–308. doi:10.1080/1369118X.2020.1863445.

Nakkeeran, N. (2010) 'Knowledge, truth, and social reality: an introductory note on qualitative research', *Indian Journal of Community Medicine: Official Publication of Indian Association of Preventive & Social Medicine*, 35(3), pp. .379–381. doi:10.4103/0970-0218.69249.

Nandutu, I., Atemkeng, M. and Okouma, P. (2021) 'Integrating AI ethics in wildlife conservation AI systems in South Africa: a review, challenges, and future research agenda', *AI & Society*, pp. 1–13. doi:10.1007/s00146-021-01285-y.

Natale, S. and Ballatore, A. (2020) 'Imagining the thinking machine: technological myths and the rise of artificial intelligence', *Convergence*, 26(1), pp. 3–18. doi:10.1177/1354856517715164.

Naudé, P. (2019) 'Decolonising knowledge: can Ubuntu ethics save us from coloniality? (Ex Africa Semper Aliquid Novi?)', in Jansen, J.D. (ed.) *Decolonisation in Universities: The Politics of Knowledge*. Johannesburg, South Africa: Wits University Press, pp. 217–238. doi:10.18772/22019083351.

Naudé, W. (2017) *Entrepreneurship, Education and the Fourth Industrial Revolution in Africa*, IZA Discussion paper No. 10855. Bon: Institute of Labor Economics (IZA). Available at: https://www.econstor.eu/handle/10419/170839.

Naudé, W. (2021) 'Artificial intelligence: neither utopian nor apocalyptic impacts soon', *Economics of Innovation and New Technology*, pp. 1–23. doi:10.1080/10438599.2020.1839173.

Nayebare, M. (2019) 'Artificial intelligence policies in Africa over the next five years', *XRDS: Crossroads, The ACM Magazine for Students*, 26(2), pp. 50–54. doi:10.1145/3368075.

Ndemo, B. and Weiss, T. (2017) 'Making sense of Africa's emerging digital transformation and its many futures', *Africa Journal of Management*, 3(3–4), pp. 328–347. doi:10.1080/23322373.2017.1400260.

Ndonye, M.M. (2019) 'Mass-mediated feminist scholarship failure in Africa: normalised body-objectification as artificial intelligence (AI)', *Editon Consortium Journal of Media and Communication Studies*, 1(1), pp. 1–8.

Nedelkoska, L. and Quintini, G. (2018) *Automation, Skills Use and Training*, OECD Social, Employment and Migration Working Papers No. 202. Paris, France: Organisation for Economic Cooperation and Development. doi:10.1787/2e2f4eea-en.

Neklason, A. (2019) 'Flashback: elite-college admissions were built to protect privilege', *The Atlantic*, 18 March. Available at: https://www.theatlantic.com/education/archive/2019/03/history-privilege-elite-college-admissions/585088/.

Neri, H. and Cozman, F. (2020) 'The role of experts in the public perception of risk of artificial intelligence', *AI & Society*, 35, pp. 663–673. doi:10.1007/s00146-019-00924-9.

Neubert, M.J. and Montañez, G.D. (2020) 'Virtue as a framework for the design and use of artificial intelligence', *Business Horizons*, 63(2), pp. 195–204. doi:10.1016/j.bushor.2019.11.001.

Neudert, L.M., Knuutila, A. and Howard, P.N. (2020) *Global Attitudes Towards AI, Machine Learning & Automated Decision Making*, Working Paper. Oxford, UK: Oxford Commission on AI and Good Governance. Available at: https://www.oii.ox.ac.uk/news/releases/global-public-opinion-split-on-benefits-ofai-finds-new-oxford-study/.

News 24 report. (2020) 'News24 to use artificial intelligence in moderating comments', *News24*, 15 July 2020. Available at: https://www.news24.com/news24/southafrica/news/news24-to-use-artificial-intelligence-in-moderating-comments-20200715.

Ng, W. (2012) 'Can we teach digital natives digital literacy?', *Computers & Education*, 59(3), pp. 1065–1078. doi:10.1016/j.compedu.2012.04.016.

Njenga, J. (2018) 'Digital literacy: the quest of an inclusive definition', *Reading & Writing*, 9(1), pp. 1–7.

Njuguna, J. (2021) 'Constructions of moral values in reader comments of the Samantha sex robot discourse in East African newspapers', *Journal of Religion, Media and Digital Culture*, 10(3), pp. 382–403. doi:10.1163/21659214-bja10013.

Novitske, L. (2018) 'The AI invasion is coming to Africa (and it's a good thing)', *Stanford Social Innovation Review*, 12 February. Available at: https://ssir.org/articles/entry/the_ai_invasion_is_coming_to_africa_and_its_a_good_thing.

Nwokoye, C.H., Okeke, V.O., Roseline, P. and Okoronkwo, E. (2022) 'The mythical or realistic implementation of AI-powered driverless cars in Africa: a review of challenges and risks', in Zgang, Y.D., Senjyu, T., So-In, C. and Joshi, A. (eds.) *Smart Trends in Computing and Communications. Lecture Notes in Networks and Systems*, Volume 286. Singapore: Springer, pp. 685–695. doi:10.1007/978-981-16-4016-2_65.

Nyetanyane, J. and Masinde, M. (2020) 'Integration of indigenous knowledge, climate data, satellite imagery and machine learning to optimize cropping decisions by small-scale farmers. A case study of uMgungundlovu District Municipality, South Africa', in Thorn, J., Gueye, A. and Hejnowicz, A. (eds.) *Innovations and Interdisciplinary Solutions for Underserved Areas. InterSol 2020. Lecture Notes of the Institute for Computer Sciences, Social Informatics and Telecommunications Engineering, Volume 321*. Springer, Cham, Switzerland, pp. 3–19. doi:10.1007/978-3-030-51051-0_1.

Obermeyer, Z., Powers, B., Vogeli, C. and Mullainathan, S. (2019) 'Dissecting racial bias in an algorithm used to manage the health of populations', *Science*, 366(6464), pp. 447–453. doi:10.1126/science.aax2342.

Obioha, U.P. and Okaneme, G. (2017) 'African humanism as a basis for social cohesion and human well-being in Africa', *International Journal of Humanities, Social Sciences and Education (IJHSSE)*, 4(5), pp. 43–50. doi:10.20431/2349-0381.040.

OECD. (2018) *Bridging the Digital Gender Divide: Include, Upskill, Innovate*. OECD Report, Paris. Available at: https://www.oecd.org/internet/bridging-the-digital-gender-divide.pdf.

OECD. (2019) *Recommendation of the Council on Artificial Intelligence*. OECD Legal Instruments. Available at: https://legalinstruments.oecd.org/en/instruments/OECD-LEGAL-0449.

Ogungbure, A.A. (2013) 'African indigenous knowledge: scientific or unscientific?', *Inkanyiso: Journal of Humanities and Social Sciences*, 5(1), pp. 12–20.

Ogunyemi, K. (2020) 'Virtue ethics traditions in Africa: an introduction', in Ogunyemi, K. (ed.) *African Virtue Ethics Traditions for Business and Management*. UK and USA: Edward Elgar Publishing, pp. 1–12.

Ormond, E. (2019) 'The ghost in the machine: the ethical risks of AI', *The Thinker*, 83(1), pp. 4–11.

Orife, I., Kreutzer, J., Sibanda, B., Whitenack, D., Siminyu, K., Martinus, L., Ali, J.T., Abbott, J., Marivate, V., Kabongo, S. and Meressa, M. (2020) 'Masakhane-machine translation For Africa'. arXiv preprint arXiv:2003.11529.

Ostherr, K. (2018) 'For tech companies, "Humanism" is an empty buzzword. It doesn't have to be', *The Washington Post*, 20 June. Available at: https://www.washingtonpost.com/news/posteverything/wp/2018/06/20/for-tech-companies-humanism-is-an-empty-buzzword-it-doesnt-have-to-be/.

Ouchchy, L., Coin, A. and Dubljević, V. (2020) 'AI in the headlines: the portrayal of the ethical issues of artificial intelligence in the media', *AI & Society*, 35(4), pp. 927–936. doi:10.1007/s00146020-00965-5.

Owe, A. and Baum, S.D. (2021) 'Moral consideration of nonhumans in the ethics of artificial intelligence', *AI and Ethics*, pp.1–12. doi:10.1007/s43681-021-00065-0.

Owoyemi, A., Owoyemi, J., Osiyemi, A. and Boyd, A. (2020) 'Artificial intelligence for healthcare in Africa', *Frontiers in Digital Health*, 2(6), pp. 1–5. doi:10.3389/fdgth.2020.00006.

Oyedemi, T.D. (2020) 'The theory of digital citizenship', in Servaes, J. (ed.) *Handbook of Communication for Development and Social Change*. Singapore: Springer, pp. 237–255. https://doi.org/10.1007/978-981-10-7035-8.

Oyowe, O.A. and Yurkivska, O. (2014) 'Can a communitarian concept of African personhood be both relational and gender-neutral?', *South African Journal of Philosophy*, 33(1), pp. 85–99. doi:10.1080/02580136.2014.892682.

Pade-Khene, C., Thinyane, H. and Machiri, M. (2017) 'Building foundations before technology: an operation model for digital citizen engagement in resource constrained contexts', in Rouco, J.C.D. and Borges, V. (eds.) *Proceedings of the 17th European Conference on Digital Government*'. Lisbon: Academic Conferences Limited, pp. 118–126.

Pagallo, U. (2016) 'Even angels need the rules: AI, roboethics, and the law', in Kaminka, G.A., Fox, M., Bouquet, P. Hüllermeier, E., Dignum, V., Dignum, F. and Van Harmelen, F. (eds.) *ECAI 2016, 22nd European Conference on Artificial Intelligence, Including PAIS 2016, Prestigious Applications of Artificial Intelligence*. The Hague, The Netherlands: IOS Press, pp. 209–215. doi:10.3233/978-1-61499-672-9-209.

Palmer, M.S., Huebner, S.E., Willi, M., Fortson, L. and Packer, C. (2021) 'Citizen science, computing, and conservation: how can "crowd AI" change the way we tackle large-scale ecological challenges?', *Human Computation*, 8(2), pp. 54–75. doi:10.15346/hc.v8i2.123.

Pangrazio, L. and Sefton-Green, J. (2021) 'Digital rights, digital citizenship and digital literacy: what's the difference?', *Journal of New Approaches in Educational Research*, 10(1), pp. 15–27. doi:10.7821/naer.2021.1.616.

Paraskeva, J.M. (2011) *Conflicts in Curriculum Theory: Challenging Hegemonic Epistemologies*. New York, NY: Palgrave Macmillan.

Park, E. (2020) '"Human ATMs": M-Pesa and the expropriation of affective work in Safaricom's Kenya', *Africa*, 90(5), pp. 914–933. doi:10.1017/S0001972020000649.

Parschau, C. and Hauge, J. (2020) 'Is automation stealing manufacturing jobs? Evidence from South Africa's apparel industry', *Geoforum*, 115, pp. 120–131. doi:10.1016/j.geoforum.2020.07.002.

Patin, B., Sebastian, M., Yeon, J., Bertolini, D. and Grimm, A. (2021) 'Interrupting epistemicide: a practical framework for naming, identifying, and ending epistemic injustice in the information professions', *Journal of the Association for Information Science and Technology*, 72, pp. 1306–1318. doi:10.1002/asi.24479.

Pellegrino, E.D. (1989) 'Character, virtue and self-interest in the ethics of the professions', *Journal of Contemporary Health Law & Policy (1985–2015)*, 5(1), pp. 53–73.

Peters, D., Hansen, S., McMullan, J., Ardler, T., Mooney, J. and Calvo, R.A. (2018) '"Participation is not enough": towards indigenous-led co-design', in McKay, D., Choi, J., Lugmayr, A., Billinghurst, M., Kelly, R., Buchanan, G. and Stevenson, D. (eds.) *Proceedings of the 30th Australian Conference on Computer-human Interaction*. New York, NY: ACM, pp. 97–101. doi:10.1145/3292147.3292178.

Phillips, T.B., Ballard, H.L., Lewenstein, B.V. and Bonney, R. (2019) 'Engagement in science through citizen science: moving beyond data collection', *Science Education*, 103(3), pp. 665–690. doi:10.1002/sce.21501.

Pietersen, H.J. (2005) 'Western humanism, African humanism and work organizations', *SA Journal of Industrial Psychology*, 31(3), pp. 54–61.

Pillay, N. (2020) *Artificial Intelligence for Africa: An Opportunity for Growth, Development, and Democratization*. South Africa: University of Pretoria. Available at: https://www.up.ac.za/media/shared/7/ZP_Files/ai-for-africa.zp165664.pdf.

Pinn, A.B. (ed.) (2021) *The Oxford Handbook of Humanism*. New York, NY: Oxford University Press.

Png, M.T. (2022) 'At the tensions of South and North: critical roles of global south stakeholders in AI governance', *2022 ACM Conference on Fairness, Accountability, and Transparency*, pp. 1434–1445. doi:10.1145/3531146.3533200.

Poisat, P., Calitz, A.P. and Cullen, D.M. (2021) 'Ethical implications and challenges of AI and RPA on the South African workforce'. Paper presented at the *20th Annual BEN-Africa Conference*, November 2021. Swakopmund, Namibia.

Polyakova, A. and Meserole, C. (2019) *Exporting Digital Authoritarianism: The Russian and Chinese Models, Policy Brief, Democracy and Disorder Series* (Washington, DC: Brookings, *2019)*, pp. 1–22.

Posada, J. (2022) 'Embedded reproduction in platform data work', *Information, Communication & Society*, 25(6), pp. 816–834. doi:10.1080/1369118X.2022.2049849.

Prisecaru, P. (2016) 'Challenges of the Fourth Industrial Revolution', *Knowledge Horizons. Economics*, 8(1), pp. 57–62.

Rafapa, L.J. (2005) 'The representation of African humanism in the narrative writings of Es'kia Mphahlele'. Unpublished Doctoral dissertation, Stellenbosch, South Africa: University of Stellenbosch, South Africa.

Rahm, L. (2018) 'The ironies of digital citizenship', *Digital Culture & Society*, 4(2), pp. 39–62. doi:10.14361/dcs-2018-0204.

Rahwan, I. (2018) 'Society-in-the-loop: programming the algorithmic social contract', *Ethics and Information Technology*, 20(1), pp. 5–14. doi:10.1007/s10676-017-9430-8.

Rani, U. and Furrer, M. (2021) 'Digital labour platforms and new forms of flexible work in developing countries: algorithmic management of work and workers', *Competition & Change*, 25(2), pp. 212–236. doi:10.1177/1024529420905187.

Rapanyane, M.B. and Sethole, F.R. (2020) 'The rise of artificial intelligence and robots in the 4th Industrial Revolution: implications for future South African job creation', *Contemporary Social Science*, 15(4), pp. 489–501. doi:10.1080/21582041.2020.1806346.

Raschka, S. and Mirjalili, V. (2019) *Python Machine Learning: Machine Learning and Deep Learning with Python, Scikit-learn, and TensorFlow 2* (3rd ed.). Birmingham, UK: Packt Publishing.

Reidsma, M. (2016) 'Algorithmic bias in library discovery systems', *Matthew Reidsma*, 11 March. Available at: https://matthew.reidsrow.com/articles/173.

Resseguier, A. and Rodrigues, R. (2020) 'AI ethics should not remain toothless! A call to bring back the teeth of ethics', *Big Data & Society*, 7(2), pp. 1–5. https://matthew.reidsrow.com/articles/173.

Richardson, B.J., Mgbeoji, I. and Botchway, F. (2006) 'Environmental law in post-colonial societies: aspirations, achievements, and limitations', in Richardson, B.J. and Wood, S. (eds.) *Environmental Law for Sustainability: A Reader*. Oxford and Portland: Hart Publishing, pp. 419–421.

Rindermann, H. (2018) *Cognitive Capitalism: Human Capital and the Wellbeing of Nations*. Cambridge, UK: Cambridge University Press.

Rios, J.A., Ling, G., Pugh, R., Becker, D. and Bacall, A. (2020) 'Identifying critical 21st-century skills for workplace success: a content analysis of job advertisements', *Educational Researcher*, 49(2), pp. 80–89. doi:10.3102/0013189X19890600.

Risberg, P. (2019) 'The give-and-take of BRI in Africa', *New Perspectives in Foreign Policy*, 17, pp. 43–47.

Rizk, N. (2020) 'AI and gender in Africa', *Open AIR*, 2 September. Available at: https://matthew.reidsrow.com/articles/173 https://openair.africa/ai-and-gender-in-africa/.

Roberts, T. and Ali, A.M. (2021) 'Opening and closing civic space in Africa: an introduction to the ten digital rights landscape reports', in Roberts, T. (ed.) *Digital Rights in Closing Civic Space: Lessons from Ten African Countries*. Brighton, UK: Institute of Development Studies, pp. 9–42. doi:10.19088/IDS.2021.003.

Roche, C., Lewis, D. and Wall, P.J. (2021) 'Artificial intelligence ethics: an inclusive global discourse?', in Masiero, S. and Nielsen, P. (eds.) *Proceedings of the 1st Virtual Conference on Implications of Information and Digital Technologies for Development (IFIP 9.4)*. Norway: Department of Informatics, University of Oslo, pp. 643–658. arXiv preprint arXiv:2108.09959.

Roff, H.M. (2019) 'Artificial intelligence: power to the people', *Ethics & International Affairs*, 33(2), pp. 127–140. doi:10.1017/S0892679419000121.

Rudschies, C., Schneider, I. and Simon, J. (2020) 'Value pluralism in the AI ethics debate: different actors, different priorities', *The International Review of Information Ethics*, 29, pp. 1–15.

Russell, S. and Norvig, P. (2010) *Artificial Intelligence: A Modern Approach* (3rd ed.). Upper Saddle River, NJ: Pearson.

Russon, M.-A. (2019) 'The push towards artificial intelligence in Africa', *BBC News*, 28 May. Available at: https://www.bbc.com/news/business-48139212.

Rutenberg, I., Gwagwa, A. and Omino, M. (2021) 'Use and impact of artificial intelligence on climate change adaptation in Africa', in Leal, Filho W., Oguge, N., Ayal, D., Adeleke, L. and da Silva, I. (eds.) *African Handbook of Climate Change Adaptation*. Cham, Switzerland: Springer, pp. 1107–1126. doi:10.1007/978-3-030-45106-

Sakellariouv, A.M. (2015) 'Virtue ethics and its potential as the leading moral theory', *Discussions*, 12(1). Available at: http://www.inquiriesjournal.com/a?id=1385.

Sako, M. (2020) 'Artificial intelligence and the future of professional work', *Communications of the ACM*, 63(4), pp. 25–27. doi:10.1145/3382743.

Saleh, M. (2022) 'Unemployment rate in Africa 2022, by country', *Statista*, 3 August. Available at: https://www.statista.com/statistics/1286939/unemployment-rate-in-africa-by-country/.

Sallstrom, L., Morris, O.L.I.V.E. and Mehta, H. (2019) 'Artificial intelligence in Africa's healthcare: ethical considerations', *ORF Issue Brief*, (312). Available at: https://www.orfonline.org/wp-content/uploads/2019/09/ORF_Issue_Brief_312_AI-Health-Africa.pdf.

Sambala, E.Z., Cooper, S. and Manderson, L. (2020) 'Ubuntu as a framework for ethical decision making in Africa: responding to epidemics', *Ethics & Behavior*, 30(1), pp. 1–13. doi:10.1080/10508422.2019.1583565.

Sander-Staudt, M. (2006) 'The unhappy marriage of care ethics and virtue ethics', *Hypatia*, 21(4), pp. 21–39. doi:10.1111/j.1527-2001.2006.tb01126.x.

Sandler, R. (2007) *Character and Environment: A Virtue-Oriented Approach to Environmental Ethics*. New York, NY: Columbia University Press.

Sandler, R. (2018) *Environmental Ethics: Theory in Practice*. New York, NY: Oxford University Press.

Santos, I. and Rubiano-Matulevich, E. (2019) 'Minding the gender gap in training in Sub-Saharan Africa: five things to know', *World Bank Blogs*, 12 August. Available at: https://blogs.worldbank.org/africacan/minding-gender-gap-training-sub-saharan-africa-five-things-know.

Sawahel, W. (2019) 'First AI faculty unveiled at global forum meeting', *University World News*, 5 April. Available at: https://www.universityworldnews.com/post.php?story=20190405064313832.

Schank, R. (2014) '"Hawking is afraid of AI without having a clue what it is: don't worry Steve"', *Roger Schank*, 8 December. Available at: https://educationoutrage.blogspot.com/2014/12/hawking-is-afraid-of-ai-without-having.html.

Schoeman, W., Moore, R., Seedat, Y. and Chen, J.Y.J. (2021) *Artificial Intelligence: Is South Africa ready?*. Accenture – Gordon Institute of business Science: University of Pretoria. Available at: https://repository.up.ac.za/bitstream/handle/2263/82719/Schoeman_Artificial.pdf?sequence=1&isAllowed=y.

Sefotho, M.M. (2021) 'Basotho ontology of disability: an afrocentric onto-epistemology', *Heliyon*, 7(3), e06540. doi:10.1016/j.heliyon.2021.e06540.

Sey, A., Razzano, G., Rens, A. and Ahmed, S. (2021) 'Mapping policy and capacity for artificial intelligence for development in Africa', *Research ICT Africa*. Available at: https://idl-bnc-idrc.dspacedirect.org/bitstream/handle/10625/60522/f191f986-416c-47c1-a06b-c1b830b735d1.pdf.

Sharda, R., Delen, D. and Turban, E. (2020) *Analytics, Data Science, & Artificial Intelligence: Systems for Decision Support* (11th ed.). Hoboken, NJ: Pearson.

Sharma, A., Jain, A., Gupta, P. and Chowdary, V. (2020) 'Machine learning applications for precision agriculture: a comprehensive review', *IEEE Access*, 9, pp. 4843–4873. doi:10.1109/ACCESS.2020.3048415.

Shehu, S. (2017) 'Nigeria imports 90% of its technology needs', *African Business Convention*, 7 February. Available at: https://businessday.ng/technology/article/nigeria-imports-90-technology-needs/.

Sheppard, A., Lewis, J., Moreu, M., Laws, M. and Hoyte, S. (2020) 'Indigenous-led technology solutions can boost biodiversity and ensure human rights (Commentary)', *Mongabay*, 27 July. Available at: https://news.mongabay.com/2020/07/indigenous-led-technology-solutions-can-boost-biodiversity-and-ensure-human-rights-commentary/.

Shevlin, H., Vold, K., Crosby, M. and Halina, M. (2019) 'The limits of machine intelligence: despite progress in machine intelligence, artificial general intelligence is still a major challenge', *EMBO Reports*, 20(10), p. e49177. doi:10.15252/embr.201949177.

Shikali, C.S. and Mokhosi, R. (2020) 'Enhancing African low-resource languages: Swahili data for language modelling', *Data in Brief*, 31, p. 105951. doi:10.1016/j.dib.2020.105951.

Shukra, Z.A., Zhou, Y. and Wang, L. (2021) 'An adaptable conceptual model for construction technology transfer: the BRI in Africa, the case of Ethiopia', *Sustainability*, 13(6), pp. 1–19. doi:10.3390/su13063376.

Sibal, P., Neupane, B. and Orlic, D. (2021) *Artificial Intelligence Needs Assessment Survey in Africa*. Paris, France: United Nations Educational, Scientific and Cultural Organization. Available at: https://ircai.org/wp-content/uploads/2021/03/UNESCO_AI_Needs_Assessment_ENG.pdf.

Sibanda, M.J. (2021) 'Academics' conceptions of higher education decolonisation', *South African Journal of Higher Education*, 35(3), pp. 182–199. doi:10.20853/35-3-3935.

Sicat, M., Xu, A., Mehetaj, E., Ferrantino, M. and Chemuta, V. (2020) 'Leveraging ICT technologies in closing the gender gap', *World Bank*, pp. 1–45. doi:10.1596/33165.

Singh, J.A. (2019) 'Artificial intelligence and global health: opportunities and challenges', *Emerging Topics in Life Sciences*, 3(6), pp. 741–746. doi:10.1042/ETLS20190106.

Siyonbola, L. (2021) 'A brief history of artificial intelligence in Africa', *noirpress*. Available at: https://noirpress.org/a-brief-history-of-artificial-intelligence-in-africa/ (Accessed: 26 July 2021).

Slote, M. (2021) 'Agent-based virtue ethics', in Halbig, C. and Timmermann, F.U. (eds.) *Handbuch Tugend und Tugendethik*. Wiesbaden: Springer VS, pp. 1–10. https://doi.org/10.1007/978-3-658-24467-5_24-1.

Small, R. (2019) 'The future of accounting: what to expect in the next five years', *South African Institute of Professional Accountants*, 11 March. Available at: https://www.saipa.co.za/future-of-accounting-what-to-expect-next-5-years/.

Smith, M. and Neupane, S. (2018) 'Artificial intelligence and human development: toward a research agenda', *International Development Research Centre* (2018) Available at: https://idl-bnc-idrc.dspacedirect.org/handle/10625/56949.

Sokona, D. (2020) 'Gendered assessment of science, technology and innovation ecosystem: case study of agricultural research and training institutions in Mali', *African Journal of Rural Development*, 5(1), pp. 63–78.

Solem, J.E. (2012) *Programming Computer Vision with Python: Tools and Algorithms for Analyzing Images* (1st ed.). Sebastopol, CA: O'Reilly Media, Inc.

South African Human Rights Commission. (2017/2018) *Equality Report 2017/2018*. Available at: https://www.sahrc.org.za/home/21/files/SAHRC%20Equality%20Report%202017_18.pdf.

Srnicek, N. (2022) 'Data, compute, labour', in Graham, M. and Ferrari, F. (eds.) *Digital Work in the Planetary Market*. Canada: The MIT Press, pp. 241–262. doi:10.7551/mitpress/13835.001.0001.

Stahl, B.C. (2021) 'Concepts of ethics and their application to AI', *Artificial Intelligence for a Better Future: An Ecosystem Perspective on the Ethics of AI and Emerging Digital Technologies*, pp. 19–33. doi:10.1007/978-3-030-69978-9_3.

Standing, G. (2014) *The Precariat: The New Dangerous Class, Trade*. London and New York: Bloomsbury.

Stats SA. (2022) *Quarterly Labour Force Survey Quarter 1: 2022*. Pretoria: Stats SA.

Stauffer, B. (2020) 'Stopping killer robots: country positions on banning fully autonomous weapons and retaining human control', *Human Rights Watch*, 10 August. Available at: https://www.hrw.org/report/2020/08/10/stopping-killer-robots/country-positions-banning-fully-autonomous-weapons-and.

Steen, M., Sand, M. and Van de Poel, I. (2021) 'Virtue ethics for responsible innovation', *Business and Professional Ethics Journal*, pp. 1–27. doi:10.5840/bpej2021319108.

Sternberg, R.J., Grigorenko, E.L. and Kidd, K.K. (2005) 'Intelligence, race, and genetics', *American Psychologist*, 60(1), pp. 46–59. doi:10.1037/0003-066X.60.1.46.

Stolper, C.D., Lee, B., Henry Riche, N. and Stasko, J. (2016) *Emerging and Recurring Data-Driven Storytelling Techniques: Analysis of a Curated Collection of Recent Stories*. MSR-TR-2016-14. Microsoft Research 2016.

Subasi, A. (2020) *Practical Machine Learning for Data Analysis Using Python*. Cambridge, MA: Academic Press. doi:10.1016/C2019-0-03019-1.

Suerdem, A. and Akkilic, S. (2021) 'Cultural differences in media framing of AI', in Schiele, B., Liu, X. and Bauer, M.W. (eds.) *Science Cultures in a Diverse World: Knowing, Sharing, Caring*. Singapore: Springer, pp. 185–207. doi:10.1007/978-981-16-5379-7_10.

Svensson, J. (2021) 'Artificial intelligence is an oxymoron', *AI & Society*, pp. 1–10. doi:10.1007/s00146-021-01311-z.

Svetelj, T. (2014) 'Universal humanism – a globalization context is the classroom of unheard options… how to become more human. The person and the challenges', *The Journal of Theology, Education, Canon Law and Social Studies Inspired by Pope John Paul II*, 4(1), pp. 23–36.

Tan, M. (2021) 'Humanistic goals for science education: STEM as an opportunity to reconsider goals for education'. In Berry, A., Buntting, C., Corrigan, D., Gunstone, R. and Jones, A. (eds.) *Education in the 21st Century*. Cham, Switzerland: Springer, pp. 159–176.

Tao, Y. and Varshney, K.R. (2021) 'Insiders and outsiders in research on machine learning and society'. arXiv preprint arXiv:2102.02279.

Terman, LM. 1916. *The Measurement of Intelligence*. Cambridge, MA: The Riverside Press.

Thatcher, J., O'Sullivan, D. and Mahmoudi, D. (2016) 'Data colonialism through accumulation by dispossession: new metaphors for daily data', *Environment and Planning D: Society and Space*, 34(6), p. 990–1006. doi:10.1177/0263775816633195.

Thakur, D., Brandusescu, A. and Nwakanma, N. (2019) '"Hello Siri, How Does the Patriarchy Influence You?": understanding artificial intelligence and gender inequality', in Sey, A. and Hafkin, N. (eds.) *Taking Stock: Data and Evidence on Gender Equality in Digital Access, Skills and Leadership*. Tokyo, Japan: United Nations University, pp. 330–338.

Thomas, A. (2011) 'Virtue ethics and an ethics of care: complementary or in conflict?', *Eidos*, (14), pp. 132–151.

Thothela, N.P., Markus, E.D., Masinde, M. and Abu-Mahfouz, A.M. (2021) 'A survey of intelligent agro-climate decision support tool for small-scale farmers: an integration of indigenous knowledge, mobile phone technology and smart sensors', in Fong, S., Dey, N. and Joshi, A. (eds.) *ICT Analysis and Applications. Lecture Notes in Networks and Systems, Volume 154*. Singapore: Springer, pp. 715–730. doi:10.1007/978-981-15-8354-4_71.

Tomašev, N., Cornebise, J., Hutter, F., Mohamed, S., Picciariello, A., Connelly, B., Belgrave, D., Ezer, D., Haert, F.C.V.D., Mugisha, F. and Abila, G. et al. (2020) 'AI for social good: unlocking the opportunity for positive impact', *Nature Communications*, 11(1), pp. 1–6. doi:10.1038/s41467-020-15871-z.

Torney, C.J., Lloyd-Jones, D.J., Chevallier, M., Moyer, D.C., Maliti, H.T., Mwita, M., Kohi, E.M. and Hopcraft, G.C. (2019) 'A comparison of deep learning and citizen science techniques for counting wildlife in aerial survey images', *Methods in Ecology and Evolution*, 10(6), pp. 779–787. doi:10.1111/2041-210X.13165.

Tschaepe, M. (2021) 'Pragmatic ethics for generative adversarial networks: coupling, cyborgs, and machine learning', *Contemporary Pragmatism*, 18(1), pp. 95–111.

Ulnicane, I., Knight, W., Leach, T., Stahl, B.C. and Wanjiku, W.G. (2021) 'Framing governance for a contested emerging technology: insights from AI policy', *Policy and Society*, 40(2), pp. 158–177. doi:10.1080/14494035.2020.1855800.

UNESCO. (2016) *A Policy Review: Building Digital Citizenship in Asia Pacific through Safe, Effective and Responsible Use of ICT*. UNESCO: Bangkok. Available at: https://smartnet.niua.org/sites/default/files/resources/246813e.pdf.

UNESCO. (2020a) *Outcome Document: First Draft of the Recommendation on the Ethics of Artificial Intelligence*. UNESCO, 7 September. Available at: https://unesdoc.unesco.org/in/rest/annotationSVC/Attachment/attach_upload_feb9258a-9458-4535-9920-fca53c95a424.

UNESCO. (2020b) *Artificial Intelligence and Gender Equality: Key Findings of UNESCO's Global Dialogue*. UNESCO, 26 August. Available at: https://en.unesco.org/system/files/artificial_intelligence_and_gender_equality.pdf.

Ungerer, M., Bowmaker-Falconer, A., Oosthuizen, C., Phehane, V. and Strever, A. (2018) 'The future of the Western Cape agricultural sector in the context of the Fourth Industrial Revolution (Synthesis Report)', *Western Cape Department of Agriculture (WCDoA), together with the University of Stellenbosch Business School (USB)*. Available at: https://www.usb.ac.za/wp-content/uploads/2018/07/THE-FUTURE-OF-THE-WC-AGRICULTURAL-SECTOR-IN-THE-CONTEXT-OF-4IR-FINAL-REP.pdf.

United Nations (UN) General Assembly. (2015) *Transforming Our World: The 2030 Agenda for Sustainable Development*, 21 October 2015, A/RES/70/1. Available at: https://www.refworld.org/docid/57b6e3e44.html.

Vallor, S. (2016) *Technology and the Virtues: A Philosophical Guide to a Future Worth Wanting*. New York, NY: Oxford University Press.

Van Breda, A.D. (2019) 'Developing the notion of Ubuntu as African theory for social work practice', *Social Work*, 55(4), pp. 439–450. doi:10.15270/52-2-762.

Van Stam, G. (2016) 'Unveiling orientalism in foreign narratives for engineering for development that target Africa', in Mawer, M. (ed.) *Development Perspectives from the SOUTH. Troubling the Metrics of [Under-]development in Africa.* Bamenda: Langaa RPCIG, pp. 197–220.

Vernon, D. (2019) 'Robotics and artificial intelligence in Africa [Regional]', *IEEE Robotics & Automation Magazine*, 26(4), pp. 131–135. doi:10.1109/MRA.2019.2946107.

VIACOMCBS. 2020. Available at: https://insights.viacomcbs.com/post/south-africans-are-optimistic-about-ai-but-unsure-of-how-it-affects-them/ (Accessed: 20 May 2020).

Vieira, L.N. (2020) 'Automation anxiety and translators', *Translation Studies*, 13(1), pp. 1–21. doi:10.1080/14781700.2018.1543613.

Vilmer, J.B.J., Escorcia, A., Guillaume, M. and Herrera, J. (2018) *Information Manipulation: A Challenge for Our Democracies.* Report by the Policy Planning Staff (CAPS) of the Ministry for Europe and Foreign Affairs and the Institute for Strategic Research (IRSEM) of the Ministry for the Armed Forces, Paris, August 2018. Available at: https://www.diplomatie.gouv.fr/IMG/pdf/information_manipulation_rvb_cle838736.pdf.

Vivienne, S., McCosker, A. and Johns, A. (2016) 'Digital citizenship as fluid interface', in McCosker, A., Vivienne, S. and Johns, A. (eds.) *Negotiating Digital Citizenship: Control, Contest and Culture.* London, UK: Rowman & Littlefield International, Ltd, pp. 1–17.

Vohland, K., Land-Zandstra Ceccaroni, L., Lemmens, R., Perelló, J., Ponti, M., Samson, R. and Wagenknecht, K. (2021) 'Editorial: The science of citizen science evolves', in Vohland, K., Land-Zandstra Ceccaroni, L., Lemmens, R., Perelló, J., Ponti, M., Samson, R. and Wagenknecht, K. (eds.) *The Science of Citizen Science.* Cham, Switzerland: Springer Nature, pp. 1–12. doi:10.1007/978-3-58278-4.

Wade, N. (2014) *A Troublesome Inheritance: Genes, Race and Human History.* New York, NY: Penguin.

Waghid, Y. and Smeyers, P. (2012) 'Reconsidering Ubuntu: on the educational potential of a particular ethic of care', *Educational Philosophy and Theory*, 44(S2), pp. 6–20. doi:10.111/j.1469-5812.2011.00792.x.

Wagner, B. (2018) 'Ethics as an escape from "Ethics-washing" to "Ethics-shopping"?', in Bayamlioğlu, E., Baraliuc, I., Janssens, L. and Hildebrandt, M. (eds.) *Being Profiled: Cogitas Ergo Sum: 10 Years of Profiling the European Citizen.* Amsterdam: Amsterdam University Press, B.V., pp. 84–89. doi:10.2307/j.ctvhrd092.18.

Wairegi, A., Omino, M. and Rutenberg, I. (2021) 'AI in Africa: framing AI through an African lens', *Communication, Technologies et Développement*, (10), pp. 1–16. doi:10.4000.ctd.4775.

Walker, D.W., Smigaj, M. and Tani, M. (2021) 'The benefits and negative impacts of citizen science applications to water as experienced by participants and communities', *Wiley Interdisciplinary Reviews: Water*, 8(1), p. e1488. doi:10.1002/wat2.1488.

Walsh, T. (2018) 'Expert and non-expert opinion about technological unemployment', *International Journal of Automation and Computing*, 15(5), pp. 637–642. doi:10.1007/s11633-018-1127-x.

Walz, A. and Firth-Butterfield, K. (2019) 'AI governance: a holistic approach to implement ethics into AI', *World Economic Forum.* Available at: https://weforum.my.salesforce.com/sfc/p/#b0000000GycE/a/0X000000cPl1/i.8ZWL2HIR_kAnvckyqVA.nVVgrWIS4LCM1ueGy.gBc.

Wang, M., Deng, W., Hu, J., Tao, X. and Huang, Y. (2019) 'Racial faces in the wild: reducing racial bias by information maximization adaptation network', *Proceedings of the IEEE/CVF International Conference on Computer Vision*, pp. 692–702.

Wareham, C.S. (2020) 'Artificial intelligence and African conceptions of personhood', *Ethics and Information Technology*, 23, pp. 1–10. doi:10.1007/s10676-020-09541-3.

Weber, F.D. and Schütte, R. (2019) 'State-of-the-art and adoption of artificial intelligence in retailing', *Digital Policy, Regulation and Governance*, 21(3), pp. 264–279. doi:10.1108/DPRG-09-2018-0050.

Weingart, P. and Meyer, C. (2021) 'Citizen science in South Africa: rhetoric and reality', *Public Understanding of Science*, 30(5), pp. 605–620. doi:10.1177/0963662521996556.

West, A. (2014) 'Ubuntu and business ethics: problems, perspectives and prospects', *Journal of Business Ethics*, 121(1), pp. 47–61. doi:10.1007/s10551-013-1669-3.

West, M., Kraut, R. and Ei Chew, H. (2019a) *I'd Blush if I Could: Closing Gender Divides in Digital Skills through Education*. EQUALS/UNESO, 2019. Available at: https://unesdoc.unesco.org/ark:/48223/pf0000367416.

West, S.M., Whittaker, M. and Crawford, K. (2019) 'Discriminating systems: gender, race and power in AI', *AI Now Institute*. Available at: https://ainowinstitute.org/discriminatingsystems.html.

White, J.M. and Lidskog, R. (2021) 'Ignorance and the regulation of artificial intelligence', *Journal of Risk Research*, pp. 1–13. doi:10.1080/13669877.2021.1957985.

Whyte, K.P. and Cuomo, C.J. (2016) 'Ethics of caring in environmental ethics', in Gardiner, S.M. and Thompson, A. (eds.) *The Oxford Handbook of Environmental Ethics*. New York, NY: Oxford University Press, pp. 234–247. doi:10.1080/13669877.2021.1957985.

Williams, G. (2019) 'Part 2: will the so-called Fourth Industrial Revolution propel SA forwards?', *Biz Community*, 14 March. Available at: https://www.bizcommunity.com/PDF/PDF.aspx?l=196&c=379&ct=1&ci=188345

Wilson-Andoh, P. (2022) 'China's belt and road initiative in Kenya', *Foreign Policy Research Institute*, 18 May. Available at: https://www.fpri.org/article/2022/05/chinas-belt-and-road-initiative-in-kenya/.

Winks, B.E. (2011) 'A covenant of compassion: African humanism and the rights of solidarity in the African charter on human and peoples' rights', *African Human Rights Law Journal*, 11(2), pp. 447–464.

Winston, A.S. (2020) 'Why mainstream research will not end scientific racism in psychology', *Theory & Psychology*, 30(3), pp. 425–430. doi:10.1177/0959354320925176.

Xi, J. (2017) 'Work together to build the silk road economic belt and the 21st century maritime silk road', *The Belt and Road Forum for International Cooperation*, 14 May. Available at: https://www.mfa.gov.cn/ce/cena/eng/sgxw/t1461872.htm. Opening speech, The Belt and Road Forum for International Cooperation, May.

Xinwa, S. (2020) 'Freedom under siege: the shrinking civic space and violations of freedom of association and assembly in sub-Saharan Africa: strategies for countering restrictions'. South Africa: Centre for the Study of Violence and Reconciliation. Available at: https://media.africaportal.org/documents/Freedom_Under_Siege_for_Web.pdf.

Yarlagadda, R.T. (2018) 'The RPA and AI automation', *International Journal of Creative Research Thoughts (IJCRT)*, pp. 2320–2882.
Yeung, K. (2019) 'Responsibility and AI', *Council of Europe Study DGI* (2019) 05. Available at: https://rm.coe.int/responsability-and-ai-en/168097d9c5.
Yew, G.C.K. (2021) 'Trust in and ethical design of carebots: the case for ethics of care', *International Journal of Social Robotics*, 13(4), pp. 629–645. doi:10.1007/s12369-020-00653-w.
Yu, H., Shen, Z., Miao, C., Leung, C., Lesser, V.R. and Yang, Q. (2018) 'Building ethics into artificial intelligence', in Lang, J. (ed.) *Proceedings of the Twenty-Seventh International Joint Conference on Artificial Intelligence (IJCAI 2018)*. Stockholm, Sweden: AAAI Press, pp. 5527–5533. doi:10.24963/ijcai.2018/779.
Yu, H.Y. (2009) 'The friendship between the care ethics and the virtue ethics', *Philosophy and Culture*, 36(2), pp. 75–92.
Zembylas, M. (2021) 'A decolonial approach to AI in higher education teaching and learning: strategies for undoing the ethics of digital neocolonialism', *Learning, Media and Technology*, pp. 1–13. doi:10.1080/17439884.2021.2010094.
Zemtsov, S. (2020) 'New technologies, potential unemployment and "nescience economy" during and after the 2020 economic crisis', *Regional Science Policy & Practice*. doi:10.1111/rsp3.12286.
Zhang, B. and Dafoe, A. (2019) *Artificial Intelligence: American Attitudes and Trends*. Oxford, UK: Center for the Governance of AI, Future of Humanity Institute. Available at: https://isps.yale.edu/sites/default/files/files/Zhang_us_public_opinion_report_jan_2019.pdf.
Zhou, J., Chen, F., Berry, A., Reed, M., Zhang, S. and Savage, S. (2020) 'A survey on ethical principles of AI and implementations', *2020 IEEE Symposium Series on Computational Intelligence*, pp. 3010–3017. doi:10.1109/SSCI47803.2020.9308437.
Zhuo, Z., Larbi, F.O. and Addo, E.O. (2021) 'Benefits and risks of introducing artificial intelligence in commerce: the case of manufacturing companies in West Africa', *Amfiteatru Economic*, 23(56), pp. 174–194. doi:10.24818/EA/2021/56/174.
Zondi, S. (2021) 'A fragmented humanity and monologues: towards a diversal humanism', in Steyn, M. and Mpofu, W. (eds.) *Decolonising the Human: Reflections from Africa on Difference and Opinion*. Johannesburg, South Africa: Wits University Press, pp. 224–242. doi:10.18772/22021036512.

Index

Page numbers in **bold** indicate tables, page numbers in *italic* indicate figures

Abebe, Rediet, 27
Abidjan, 50
Access Now, 58
Accra, 50
activism on digital platforms, 58–59
Adams, Nathan-Ross, 3
Addis Ababa, 50
Afonja, Tejumade, 13
African humanism, 1–2, 4
African Institute for Mathematical Sciences, 88
African Observatory on Responsible AI (AORAI), 89
African Union (AU), 21, 39
Afro-feminist thought, 66
Algeria, 20, **23**
algorithmic biases, 8, 35
algorithmic exploitation, 17–20
Alibaba, 50
Amazon, 18, 50, 88
Angola, 51
anthropocentrism, 1, 32
Apple, 50, 88
Aristotle, 33
Article 19, 58–59
artificial intelligence (AI), 1
  Africa-inclusive view of, 38
  African perceptions of, 46–47
  ahistorical view of, 55
  AI-centrism, 1, 32
  artificial general intelligence (AGI), 31–32, 55
  design and deployment of, 2–3, 7, 11, 37, 42, 80
  disruption of the labour market, 72–77
  diversity crisis, 8, 11, 13
  ethical debates around, 29–33
  gendered dimension, 14–17
  governance, 11, 22, 24–26
  hype, 42, 46
  inequalities and inequities, 11, 25, 27, 38, 43, 60
  moral reasoning, 30, 32
  narrow, 31
  nationalism, 22
  social good, 5, 20, 88–89
  stakeholders-in-the-loop, 35, 42, 55
  summers and winters, 55
  superpowers, 22
  value ladenness, 11
  West's domination over, 13
Artificial Intelligence for Development Africa (AI4D Africa), 22, 89
AI for Social Good (AI4SG), 88
  insidious side, 88
artificial neural network (ANN), 36, 86, 91, 92
Association for Computational Linguistics (ACL), 91
automation, 19, 27, 53, 56, 62, 71–80
  intelligent automation, 19

Bace API, 64
Baidu, 50
Bambara, 91
Bardzell, Shaowen, 42
Baum, Seth, 1
Belt and Road Initiative (BRI), 20
  Chinese BRI investment, 52
  economic impact on Africa, 52
Bentley, Peter John, 32
Benyera, Everisto, 7
Biko, Steve, 4

Birhane, Abeba, 3
black box conundrum, 36
Black Consciousness, 4
Black in AI, 5
Braga, Daniela, 27
Botswana, 5, 20, **23**, 39, 53, 93
Buolamwini, Joy, 37
Burundi, 6

Caboz, Jay, 48
Cains, Andrew, 91
Cambridge Analytica, 20
Cameroon, **23**, 58, 59
Central America, 21
Centre for Intellectual Property and Information Technology Law (CIPIT), 89
Chakanya, Naome, 19
chatbots, 53, 90
Chimedza, Tinashe, 49
China, 20, 39, 50–52
Cisse, Moustapha, 13
citizen science, 65, 67–68, 89
Cohen, Philip, 12
clickworkers, *see* ghost workers
CloudWalk, 51
colonialism, 3
  digital, 55
  exploitation, 6, 11
  neoliberal economic models, 18–19
  post-colonial abuses, 7, 50
  post-colonial political economies, 11, 16
coloniality, 34
  algorithmic, 66
  beta-testing, 19–20
  digital, 95
colonisation, 6–7
  cyber colonisation, 20, 28
Comaroff, John, 6
Comaroff, Jean, 6
communalism, 4
computer vision (CV), 83–86, 92–93
Consciencism, 4
Convention on Cyber Security and Personal Data Protection, 39
convoluted neural network (CNN), 92
Côte d'Ivoire, 20
Council for Scientific and Industrial Research (CSIR), 93
COVID-19, 51
critical itinerant curriculum, 64
critical race theory, 11–12
critical science approach, 7
Crosston, Matthew, 20
Curiosity, 92
cyber-activism, 59, 61

Dakar, 50
Data for Black Lives, 5
data imperialism, 22
decolonisation of AI, 66
  ethical component, 67
  of policy discussions around ethical AI, 11
  of power, 5
  structural, 6–7
Deep Blue, 83
deep feedforward network, 86
deep learning, 83, 84, 86–90
Deep Learning Indaba, 5, 88
Democratic Republic of the Congo, 5, 93
digital access, 60–62
digital authoritarianism, 51, 61
digital citizenship, 61–68
  critical, 65–68
  curtailment of, 58–59
  precarious concept, 58
  radical, 61
  rights, 59, 61, 62
digital dictatorships, 51, 57–58
digital divide, **15**, 37, 57, 95
  first-level, 37
  second-level, 37–38
  unequal digital access, 59–61
digital economy, 49, 51
digital education, 62–65
digital entrepreneurship, 50
digital infrastructure, 50
digital literacy, 60–62
  ethical, 61
disinformation, 59
divination systems, 3
Djibouti, 51, 71
drought early warning systems (DEWS), 41
Du Plessis, Gretchen, 17, 34

East Africa, 39
East Asia, 46
ECOCASH, 49
ecology, 20–21
  carbon emissions, 21
  climate change, 21, 40, 56, 64
  deforestation, 21

ecological sensitivity, 32, 40
habitat loss, 21
nutrient loading, 21
overexploitation, 21
pollution, 21
Egypt, **16**, **23**, 53, 58, 59, *87*
Elephant Listening Project (ELP), 41
enactive cognitive science, 66
ethical AI, 11
accountability, 30, 36, 42, 58, 90
advocacy, 42, 45, 57, 58, 89, 95
autonomy, 30, 31
beneficence, 8, 30
care ethics, 31, 34, 40
deontology, 29, 31
environmental virtue ethics, 33
ethical debates around, 29–33
ethics shopping, 43
ethics washing, 28
eudaimonia, 31, 32
explainability, 29, 36, 37, 78
fairness, 63, 77, 78, 81
inclusivity, 45, 50
justice, 78, 81
neo-Aristotelian virtue ethics, 31, 32
non-maleficence, 36
participation, 7, 11, 41, 42, 45, 50, 58, 60, 68, 91
privacy, 6, 8, 11, 13, 22, 30, 36, 38–40, 45–47
relational ethics of care, 21, 33–41, 45, 66, 78, 95
responsibility, 34, 59, 61
restorative justice, 39
social justice, 49, 64
'technomoral practices', 29
transparency, 8, 35, 36
utilitarianism, 29–31
virtue ethics, 29–33
in the workplace, 77–78
Ethiopia, 51, 52, 59, *87*
Eurocentric knowledge, 2
Western-centric epistemologies, 63, 65
Evans, Gavin, 12
Ewe, 91

Facebook, 17, 18, 50, 51
facial recognition software, 20, 64
feminism
ubuntu, 6, 17, 34, 38
fourth industrial revolution (4IR), 7, 14, 16, 43, 45, 50, 62, 64, 71, 95

Gabon, 50, 60
Garbe, Lise, 58
gender disparities, 6
digital exclusion, 14
historical gendered inequalities, 16
in STEM disciplines, 14
Ghana, **23**, 39, 48, 52, 53
ghost workers, 18, 19
gig economy, 18, 19, 49, 63, 78, 95
Gilligan, Carol, 33
Global North, 21, 22, 45, 62, 66
Global South, 11, 19, 21, 24, 25, 87, 89, 91
Google, 37, 50, 74, 86, 88, 91, 92
Gravett, William, 39

hashtag campaigns, 59
Hawking, Stephen, 55
humanism, 1, 2
African humanism, 4
and AI, 2–4
humanistic perspective, 1, 4
view of ethical AI governance, 25–26
Huawei, 20, 50

IBM Malaria Challenge, 88
IBM Watson, 74, 83
indigenous knowledge systems, 89
coalescing of scientific knowledge and indigenous knowledge, 64
continued illegitimacy of African knowledge systems, 63
dismissal of, 6, 21
epistemicide, 64
intelligence quotient (IQ), 12
Internet shutdowns, 58

job polarisation, 72
Johannesburg, 50, 93

#KeepItOn, 58
Kampala, 50
Kasparov, Garry, 83
Kaunda, Kenneth, 4
Kenya, 7, **15**, **16**, **23**, 39, 41, 48–52, 59, 74, *87*, 89, 91
Kigali, 50

Lagos, 50
Larson, Erik, 31
Lesotho, **23**
localised innovation, 65

Luganda, 91
Lyons, Angela, 62

Macfarlane Smith, Genevieve, 27
Machine Intelligence Institute of Africa, 88
machine learning (ML), 62, 64, 74, 79–81, 83–86
Machine Leaning Africa, 88
machine learning from text, *see* text mining
Madagascar, 60
Malanga, Donald, 47
Mali, **16**, **23**, 52, 60
Maputo, 50
Markauskaite, Lina, 80
Masakhane, 91
Mauritius, **23**, 60
Meta, *see* Facebook
Metzinger, Thomas, 28
Microsoft, 18
Molefe, Motsamai, 4
Montréal Declaration for the Responsible Development of Artificial Intelligence, 40
Morocco, 60
Mozambique, **15**, 41
M-Pesa, 49
multiplayer perceptron (MLP), *see* deep feedforward network
Munoriyarwa, Allen, 48
Musk, Elon, 55

Namibia, 7, **23**, 53
named-entity recognition (NER), 86, 90, 91
natural language processing (NLP), 64, 83, 86, 90, 91
Nayebare, Michael, 25
N'Guessan, Charlette, 64
neural machine translation (NMT), 91
Nigeria, **16**, 20, **23**, **24**, 39, 48, 50, 52, 53, 59, 74, 87
Niyel, 89
Nkrumah, Kwame, 4
non-human animals, 21, 40
  human-ecological relationships, 1
non-human entities, *see* non-human animals
North Africa, 20, 39
Nyer, Julius, 4

Outcome Document
  First Draft of the Recommendation on the Ethics of Artificial Intelligence, 40
Owe, Andrea, 1

Payne, Katy, 41
personhood, 5
  individualist aspect, 6
  relational dimension of, 6
plurality, 4, 5, 35, 45, 50, 58, 63, 95
  criticism of, 36
pluriversality, 66, 95
Pretoria, 50
pseudo-sciences, 7
  eugenics, 7, 12
  racial science, 12

racial science, *see* race science
racism, 12
  colonial, 17
  racial and intellectual superiority of white men, 7
  racial bias in deep facial recognition, 36
  racialised notion of an intelligence hierarchy, 13
  technological superiority, 7
  white racial frame, 12, 13
Rahm, Lina, 57
rationality, 38, 63
Recommendation of the Council on Artificial Intelligence, 40
Research ICT Africa, 89
responsible innovation (RI), 33
Russon, Mary-Ann, 13
Rwanda, 7, **23**, 87, 88, 91

SAFARICOM, 49
Safe city projects, 39
Sandler, Ronald, 32
Schank, Roger, 55
science, technology, engineering, and mathematics (STEM), 14, 68, 80
  humanistic STEM, 80
  STEM education, 64
scientific racism, *see* race science
Senegal, **23**, 89
Sierra Leone, **23**
Signal, 59

Slote, Michael, 31
Small, Rashied, 48
smartphones, 60
Snapshot Serengeti Challenge, 88
social robots, 54
socio-materialism, 8
socio-technical lens, 7, 25, 62
soft skills, 62
Somalia, 7
South Africa, **15**, **16**, 20, **24**, 37, 41, 46–48, 52, 53, 59, 60, 64, 71, 74, 77, *87*, 88, 93
South America, 21
South Sudan, 7, **24**
Sudan, 58, 59, 60
Sub-Saharan Africa (SSA), 3, 14, 20, 21, 46, 61, 64
surveillance technology, 20
sustainable development goals (SDGs), 60, 88
Swahili Social Media Sentiment Analysis Challenge, 91

Tanzania, **16**, 48, 49, 67, 74, *87*, 88
technology
  appropriate technology theory, 65
  emerging technologies, 5, 14, 25, 34, 50, 54, 62
  technochauvinism, 54
  technocratic approach to mitigating disparities, 37
  techno-dystopia, 49–52
  technological determinism, 28, 89
  technological infrastructure, 22, 38
  technological leapfrogging, 49
  technological oligarchs, 19, 95
  techno-utopia, *see* techno-dystopia
Telegram, 59
Tencent, 50
text analytics, *see* text mining
text mining, 81, 84, 90–93
Togo, 58
trade unions, 19
transnational digital platforms, 50
trolling activity, 51
Tsonga, 91
Tunisia, **16**, **23**, 60, *87*
Twi, 91
Twitter, 51, 91

ubuntu, 1–4, 6, 17, 25, 33, 34, 37, 38
  and AI, 4–6
Uganda, **23**, **24**, 39, 48, 59, 74, *87*, 91
Ujamaa, 4
United Nations Educational, Scientific and Cultural Organization (UNESCO), 8
United Kingdom, 20, 22
United States, 12, 20, 22, 88

Van Stam, Gertjan, 25
virtual private networks (VPNs), 59
voice assistants, 53

Wade, Nicholas, 12
Western humanism, 2–4
World Economic Forum (WEF), 24

Yaoundé, 50

Zamba, 93
Zambia, **24**, 39, 59, *87*
Zhongxing Telecommunications Equipment (ZTE), 51
Zimbabwe, 20, **23**, 49, 51, 58, 59